Other books by Richard Perry

THE WORLD OF THE TIGER

THE WORLD OF THE POLAR BEAR

THE WORLD OF THE WALRUS

THE WORLD OF THE GIANT PANDA

THE WORLD OF THE JAGUAR

THE UNKNOWN OCEAN

AT THE TURN OF THE TIDE

THE POLAR WORLDS

Illustrated by
Nancy Lou Gahan

VOLUME II: THE MANY WORLDS OF WILDLIFE SERIES

THE POLAR WORLDS

Richard Perry

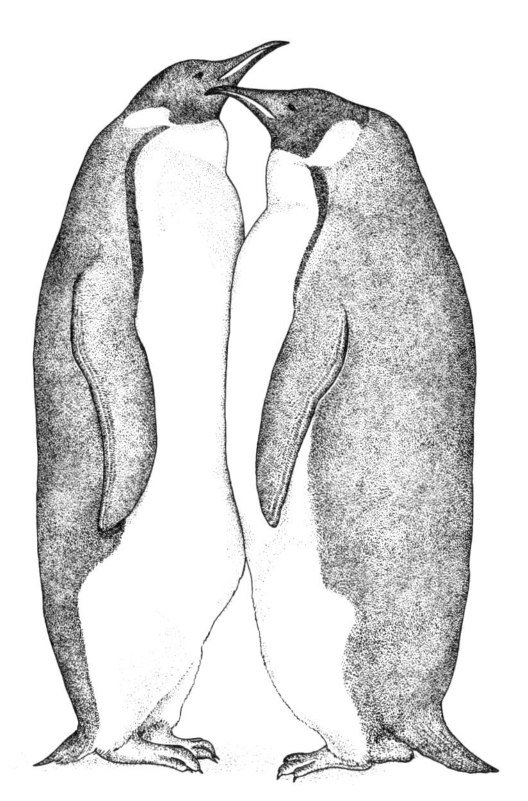

Taplinger Publishing Company New York

First Edition

Published in the United States in 1973 by
TAPLINGER PUBLISHING CO., INC.
New York, New York

Published simultaneously in the Dominion of Canada
by Burns & MacEachern Ltd., Ontario

Library of Congress Catalog Card Number:
77-179494

ISBN 0-8008-6405-0

Designed by Mollie M. Torras

A belated memorial to
the tens of thousands of Husky sledge-dogs
who have fought, starved and died
during 125 years of polar exploration

ACKNOWLEDGEMENTS

It is a pleasure to acknowledge that, in the conditions under which I work, this book could not have been written without the extraordinary efforts of the Reference Department of the Northumberland County Library in obtaining for me the necessary reference books and journals, often by way of that unique institution, the National Central Library.

I am also indebted to the Librarian of the Scott Polar Research Institute; to Sir Alistair Hardy for permission to quote from *Great Waters* (Collins, and Harper & Row, Publishers, Inc., 1967) and to Dr. Bernard Stonehouse for similar permission in respect of *The King Penguin of South Georgia* published in the Scientific Reports of the Falkland Islands Dependencies Survey; to Cassell & Company, Ltd. and Taplinger Publishing Co., Inc., who have allowed me to draw upon material used in *The World of the Walrus* (1967), and to the University of Washington Press for similar use of *The World of the Polar Bear* (1966).

CONTENTS

ILLUSTRATIONS

DRAWINGS

MAPS

Introduction

In introducing *The Unknown Ocean*, the first volume in this series, I described it as essentially a book of questions. In this second volume, too, many problems are left unresolved—inevitably, because we are dealing not only with the sea again, that difficult environment in which to work and observe, but also with the almost as difficult conditions prevailing in the Arctic and Antarctic. The latter has only been explored in any detail by zoologists during the past thirty years, while access to information on the natural history of the Arctic, so much of which lies in the Soviet sphere of influence, is still retarded by the language barrier. It is this barrier which presumably accounts for the fact that no general ecology of the Arctic exists; for instance, *The Arctic Year* by Peter Freuchen and Finn Salomensen—which I have found most helpful—is virtually restricted to Greenland. Even this work has no counterpart in the Antarctic, and for information on the natural history of that ruthless but fascinating region one must comb the books of explorers and a variety of scientific journals.

RICHARD PERRY

Northumberland, 1972

I
THE ANTARCTIC

During the noon hours the silver rays are lost,
and the moon itself is changed to a deep orange
yellow in the diffused twilight cast by the
gleaming crimson band to the north; but as the red
glow slowly travels around and is lost behind the
western hills, our white world is left alone with
the moon and the stars. The cold, white light
falls on the colder, whiter snow against which the
dark rock and intricate outlines of the ship stand
out in blackest contrast. Each sharp peak and
every object about casts a deep shadow, and is
clearly outlined against the sky. . . . The eye travels
on and on over the gleaming plains till it meets
the misty white horizon, and above and beyond,
the soft, silvery outlines of the mountains.

ROBERT FALCON SCOTT
The Voyage of the "Discovery",
McMurdo Sound, 1 August 1903

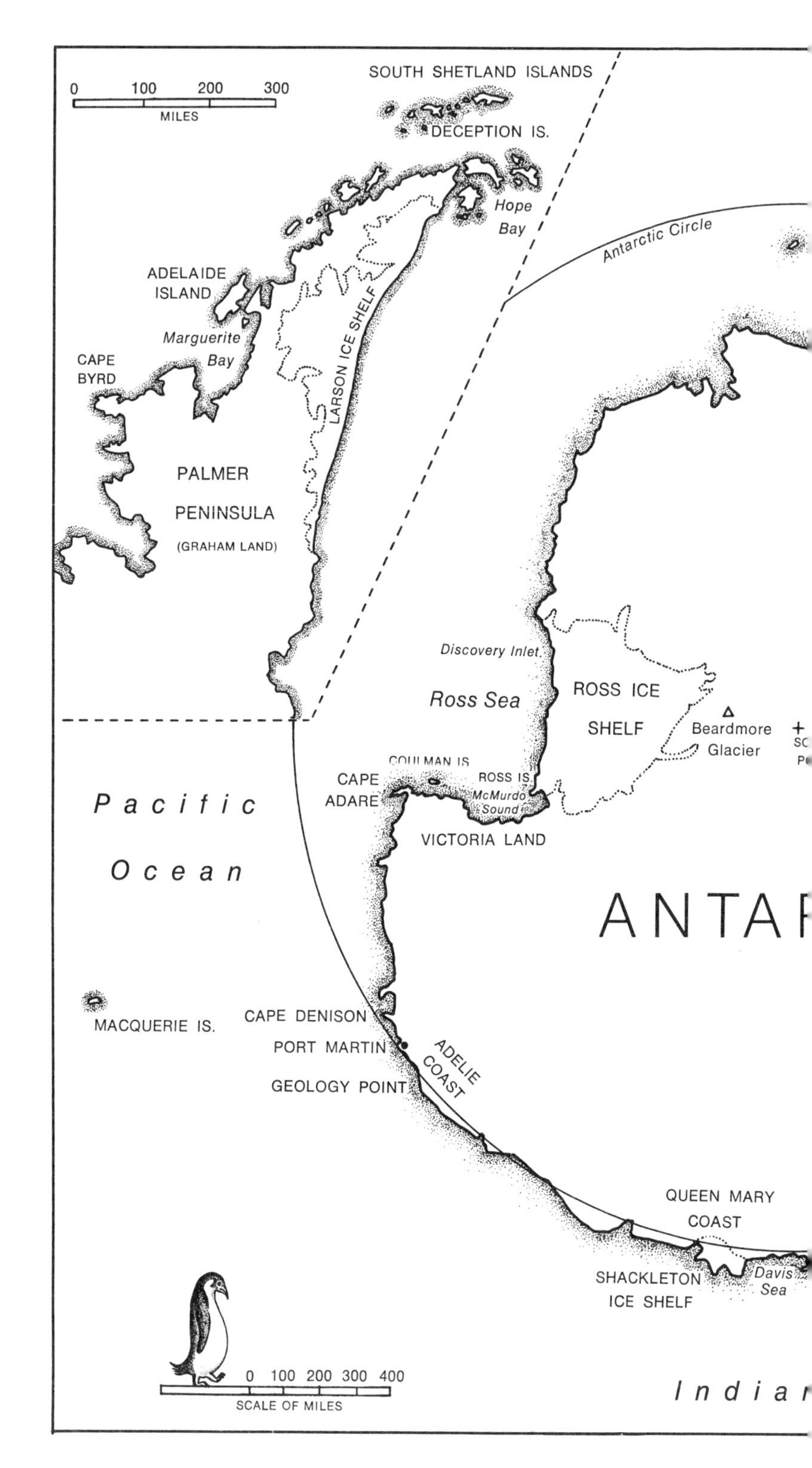

0 100 200 300
MILES
SOUTH SHETLAND ISLANDS
DECEPTION IS.
Hope Bay
Antarctic Circle
ADELAIDE ISLAND
Marguerite Bay
CAPE BYRD
LARSON ICE SHELF
PALMER PENINSULA
(GRAHAM LAND)
Discovery Inlet
Ross Sea
ROSS ICE SHELF
Beardmore Glacier
COULMAN IS.
CAPE ADARE
ROSS IS.
McMurdo Sound
VICTORIA LAND
Pacific Ocean
MACQUERIE IS.
CAPE DENISON
PORT MARTIN
GEOLOGY POINT
ADELIE COAST
QUEEN MARY COAST
SHACKLETON ICE SHELF
Davis Sea
0 100 200 300 400
SCALE OF MILES

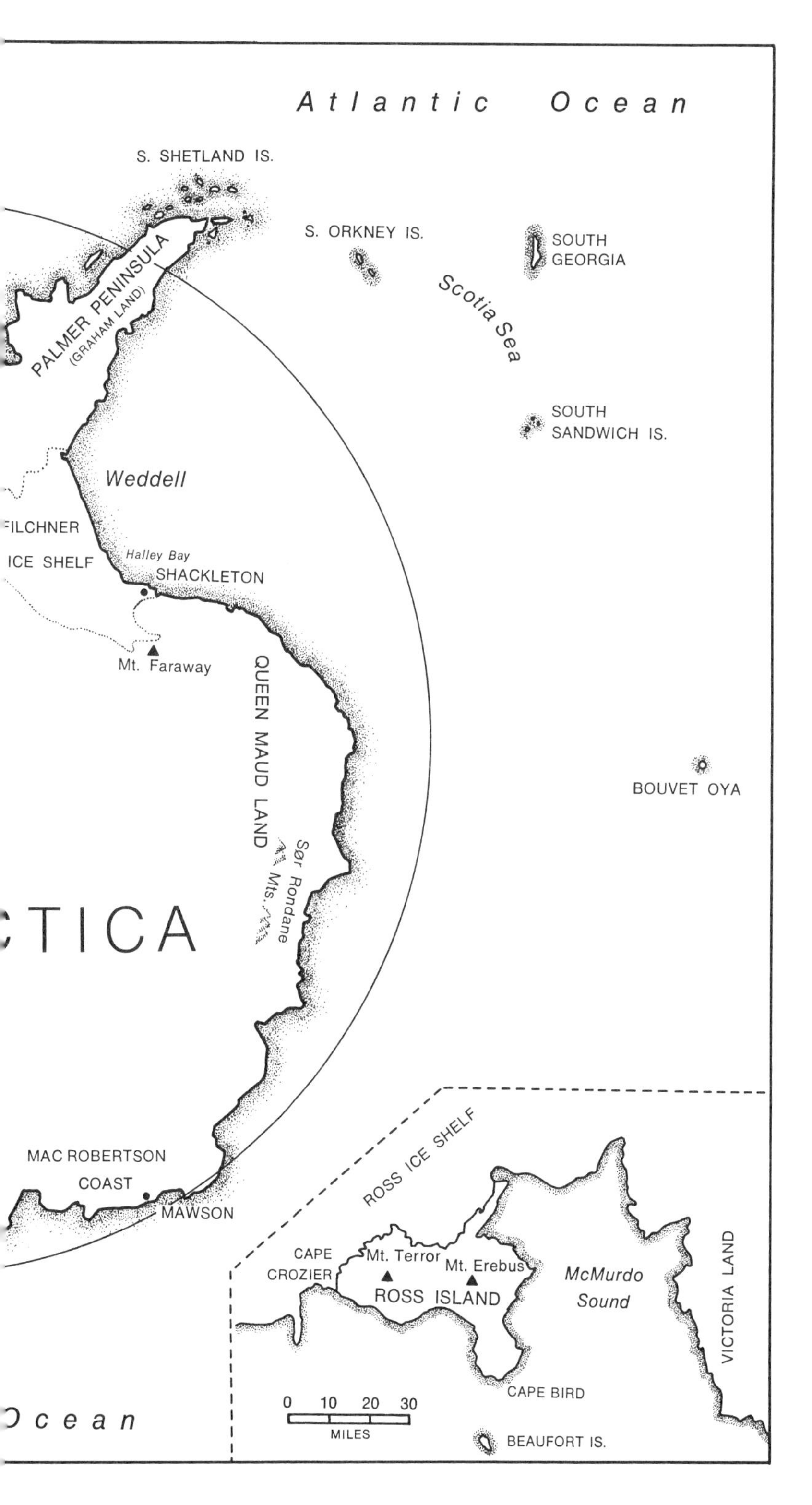

Atlantic Ocean
S. SHETLAND IS.
S. ORKNEY IS.
SOUTH GEORGIA
Scotia Sea
PALMER PENINSULA
(GRAHAM LAND)
SOUTH SANDWICH IS.
Weddell
FILCHNER
ICE SHELF
Halley Bay
SHACKLETON
Mt. Faraway
QUEEN MAUD LAND
BOUVET OYA
Sør Rondane Mts.
CTICA
MAC ROBERTSON COAST
MAWSON
ROSS ICE SHELF
CAPE CROZIER
Mt. Terror
Mt. Erebus
ROSS ISLAND
McMurdo Sound
VICTORIA LAND
CAPE BIRD
0 10 20 30
MILES
BEAUFORT IS.
Ocean

1: The Nature of the Antarctic

A continent of frozen snow as large as the USA and Mexico combined, encompassed by 12 million miles of pack-ice and frigid seas. Virtually one gigantic glacier with a 300-foot veneer of snow overlaying an ice-cap 1¼ miles thick, accumulated by millennia of compression, since the annual snowfall over the interior of Antarctica averages only 5 or 6 inches. No other continent is so uniformly high, for its plateaus lie between 6,500 and 13,000 feet above sea-level, and the rocky peaks of its great mountain ranges tower above the ice to heights of almost 17,000 feet.

There is no other large mass of land within 2,000 miles of Antarctica; and the presence in the middle of an ocean of this polar Tibet of 5½ million square miles, and in particular the summer melt from its storehouse of nine-tenths of Earth's snow and ice, and the slow flow of its glaciers over the coastal ice-shelves, determine not only the climate and composition of the Antarctic ocean, but influence those of lands and seas far beyond.

How is one to define the boundaries of the Antarctic? The Polar Circles, beloved of cartographers, are no more than an astronomical concept denoting a latitude (66° 33′) north or south of which occurs the phenomenon of the midnight sun. Such an abstract concept is of little significance to the naturalist, searching for some concrete boundary demarcating one type of environment from another. This he finds in the Antarctic Convergence, more conveniently termed the Polar Front, which is a belt of sea, varying from 20 to 30 miles in width, where two different kinds of water meet. One, a layer of warm saline water, flows down from the north; the other, a cold layer, less saline since it is diluted with fresh water from Antarctica's summer snow-melt and also from the vast quantities of pack-ice drifting in it, flows up from the Antarctic. No two oceanologists are in agreement as to precisely how these different masses of water are constituted, nor indeed how they behave when they meet; but we know that in the latitude of the Polar Front the cold layer sinks as the warm layer rises, and that there is not only a sharp rise in the sea's temperature but also a change in its chemical composition, because the ascending waters bring up from the deeps to the surface large amounts of such mineral salts as phosphates and nitrates, on which plankton, the sea's food base, thrive.

So influential is this sea-change in the belt of the Polar Front that if the small euphausian crustaceans brought up by a trawl are *Euphausia frigida* then the likelihood is that the ship is on the Antarctic side of the convergence; if they are *longirostris* or *vallentini* then it is likely to be on the northern side. Whether this be strict fact or slight exaggeration it is beyond dispute that to the north of the Front the waters of the sub-Antarctic are relatively poorly supplied with plankton, and therefore with marine fauna and avifauna; while to the south there are suddenly numbers of seabirds wheeling and turning in the almost constant mist of condensation, as the warmer air moves over the cold, teeming waters of the Antarctic, which are super-abundantly fertilized not only by the upwelling along the Front, but also by upwellings of warmer water around the coast of Antarctica. There the ice is crystallized fresh water: whereas the unfrozen sea adjacent to the ice is salt and, being heavier, sinks along the continental slope to become the Antarctic bottom current, while the fresh water flows over it eastwards around Antarctica and northwards to the Polar Front.

Although the Front does not follow any exact line of latitude, it approximates to 50° S in the Atlantic and Indian sectors of the southern ocean, passing far north of South Georgia but close to Kerguelen; and to between 55° and 62° S in the Pacific sector, where it passes close to Macquerie Island and midway between Cape Horn and the South Shetlands. There is a remarkable difference not only in the marine life, but also in the climate and in the flora and fauna of the lands and islands north of the Front and those south of it. The contours of the Front also coincide very closely with those of the mean annual isotherm of 43 degrees F, and with the northern limit of the pack-ice, though the latter varies both in configuration and latitude from year to year. The presence and extent of the pack-ice is as important a factor in determining the climate and fauna of the lands within its influence as is that of the Front.

If sudden differences in the sea's temperature and in its fauna and avifauna inform the naturalist immediately he has crossed the Polar Front into the Antarctic, the mariner will not be long behind him in detecting that he is now in his love-hate realm of the pack-ice. First he sees the "ice-blink" of light reflected from the pack on large icebergs or overcast sky, and then the pack itself—at the outset small fragments of ice; then floes 200 or 300 yards across, with a mile or more of open water between them; and soon belts of broken pack stretching away endlessly to port and starboard, dazzling white in the almost black water under the dull gray sky.

"The water appears like ink because it receives so little of the light which is nearly all reflected off the ice; it has the darkness of a cave", wrote Sir Alistair Hardy in *Great Waters* when south-east of Bouvetøya, more than 1,000 miles off the coast of Queen Maud Land:

> The whole scene is one of reversed values. The ice-floes, which take the place of land, are as brilliantly bright as the sky should be, whereas the sky itself is gray and dark in comparison and the water darker still. . . . Yet there is color within the ice. Every little crack and hollow appears as if lit by blue light: not a sky blue, an ice blue. . . . There is no blue in the sea or sky for it to be reflecting; it is the real blue of transparent ice.

It is in the Antarctic's autumn, early in February in the coldest seas, such as the Weddell Sea, but not until March or early April in more exposed regions such as the Ross Sea, that the pack-ice begins to form. When the temperature falls to 10 or 5 degrees F in still weather, the broad belt of cold water encircling Antarctica freezes rapidly. As the crystals of ice multiply in sheltered waters, so the underlying swell fuses them into pancakes up to 6 feet in diameter, resembling Victoria regina lilies with upturned edges. Snow falling on the pancakes consolidates them into solid floes, and as temperatures continue to fall and the swell continues its jostling and jamming, the floes grow into sheets of ice, 30 or 40 feet thick and 1 or 1½ miles in extent. These spread outwards during the winter and spring, and the currents carry the pack northwards away from Antarctica into the open sea. At the same time new ice is forming more slowly along the interstices in the old offshore pack-ice, which extends to join the inshore ice and form an almost continuous belt by August or September. In summer, when air and sea temperatures rise, the pack-ice begins to disintegrate along its northern edge towards the Polar Front. With cracks like rifle-shots black stripes of open water cut fissures through the floes, which open in a matter of minutes into broad leads, on whose jagged edges sunlight plays. Ice ridges, 40 or 50 feet high, collide with peals of thunder, upending huge twisted blocks, screwed up by the pressure, which crash down on to the floes, as the heavy old blue-green ice ploughs relentlessly through the newer ice, though this may be 6 feet thick. In six weeks the pack may have receded by as much as 750 miles, and by the end of January large masses of ice may be restricted to such areas as the Weddell Sea, where the pack is deflected westwards and jammed against the east coast of the 650-mile-long Palmer Peninsula (or Graham Land). There the sea is almost covered with heavy floes, many years old, which the constant pressure has piled one on top of another, and some sheltered bays remain frozen for years together.

This bay ice is not to be confused with the ice-shelves which fringe a third of Antarctica's coastline. These are either masses of land-ice, some of it afloat but still attached to the land, composed of the accumulation of layers of snow which have not been pressed to the consistency of glacier ice; or the fronts of glaciers, flowing at the rate of a few inches or feet a year down from the ice-cap and out over the sea. Some of the latter are of gigantic proportions. The Ross Ice-shelf, for example, presents at its seaward edge a barrier of ice-cliffs upwards of 200 feet high and 400 miles in length, and sweeps landwards for 350 miles to the chaos of ridges and crevasses of the Beardmore and Axel Heiberg glaciers, where it is 1,500 feet thick, and then to within 500 miles of the Pole itself. From these ice-fronts vast slabs of ice break off from time to time, and give birth to those collossal flat-topped and tabular bergs—some, a hundred miles or more in length —which are a spectacular feature of Antarctic seas.

To a naturalist, then, there are four major environments in the Antarctic: the pack-ice and the open sea; the oceanic islands and their inshore waters; the ice-shelves, coastal cliffs and headlands of Antarctica; and the interior of this vast continent. Bearing in mind the extensive snow and ice coverage, and the long and uniquely severe winters, all except the sea would appear, superficially, to be hostile to any of the higher forms of life. Yet, while it is true that there are no indigenous land mammals in the Antarctic, seals are to be numbered in millions; and there are tens of millions of sea-birds, including some small species of petrels which may be among the most numerous birds on Earth.

2: Seals of the Antarctic

The pack-ice is the home for the greater part of their lives of many of the inhabitants of the Antarctic. Seals of various species lie out on the floes, sleeping or basking in the sun. Penguins line the bergs or, shooting a foot or so out of the water between the floes, project themselves through the air in graceful streamlined arcs and re-enter the sea without causing a ripple. A few seconds later and a hundred feet ahead, they repeat the action—as smoothly and efficiently as dolphins. Whales burst through the ice-cracks to breathe, their vast curving backs and fins turning like huge wheels, the air hissing from their lungs; then, standing vertically on their tails they push their colossal heads above the ice before sinking slowly from sight. Dove-like snow petrels, so purely white that they are lost against white floes or glistening bergs, flash into view over the dark waters of leads, or hover over the edges of floes, dipping down ever and again to pick up the euphausian shrimps and tiny fish washed on to them. Sharp-winged Antarctic petrels or ice-birds—for their presence, like that of snow petrels, invariably heralds ice—skim the crests of the waves and swoop into their troughs.

But the pack-ice is an almost unexplored habitat, particularly during the winter months when thousands of square miles are inaccessible to ships. Until quite recently, for example, it had always been supposed that the curiously pug-nosed Ross seal was an extremely rare species, since fewer than fifty had been seen up to 1940. However, the introduction of powerful ice-breakers, equipped with spotter helicopters, has made it possible for research ships to venture deeper into the ice-fields; and it is now believed that there may be as many as 20,000 of these seals though obviously all estimates of seal populations in such an environment must be in the nature of guesses pleasing to zoologists' palates. No doubt the apparent rarity of Ross seals is due partly to their solitary habits—no more than six have ever been seen on one floe—and partly to the fact that they are locally distributed in the pack-ice, their exclusive home. So far as is known they never haul out on beaches, and they have never been recorded north of the pack. We do know that their hooked and recurved needle-sharp incisors and canines enable them to capture squid in particular and also fish; but we do not know anything about their social and breeding habits, nor even the size of a fully grown adult, though it, like the Weddell seal, probably reaches a length of 11 feet.

By contrast, the relatively small crab-eaters—which average from 7½ to 9 feet in length and whose weight of 500 pounds is not much more than half that of a Weddell seal—are certainly the most numerous seals in the Antarctic and possibly in any ocean, with an estimated population of between 5 and 8 million. The pack-ice is their home too. Far out in it they give birth to their pups in the spring and early summer, and they do not venture beyond its northern limits. However, they cannot range as far south in the winter as the Weddell seals, which mainly inhabit the shelf and bay-ice, because, unlike them, they cannot maintain breathing holes in the heavy floe-ice with their teeth.

The crab-eaters are misnamed, for they do not feed on crabs, but on those small euphausian crustacea, the krill, which swarm in the leads of open water among the ice-pack. Their tri-pointed teeth are so designed that the cusps on those in the lower jaw fill the spaces between those in the upper jaw when their mouths are closed, thus forming a perfect sieve for the retention of the krill when the water flows out of their mouths. Their teeth cannot therefore be used as grinders for mashing up the krill (or for gnawing holes in the ice) and the former operation is presumably performed by the grit which is commonly found in their stomachs and intestines.

All seals are in their element in the water, but crab-eaters are also exceptionally lithe and active in their movements when out of it, propeling themselves at speeds of 15 miles per hour over ice and snow with lateral "swimming" strokes of tails and flippers. Nevertheless, most adults and many young ones are scarred with parallel gashes, upwards of 2 feet in length, and some Ross seals bear similar wounds. Were only crab-eaters scarred in this manner one might presume killer whales to be responsible, though it is difficult to understand how such a high proportion of them can be wounded but not captured. Killers in packs of a dozen or a score or even a hundred are ever present in the pack-ice, constantly rising to blow in the leads, and probably range as far south in the Antarctic as there is broken ice; but though these highly successful and intelligent predatory dolphins are capable of shattering floes at least 2½ feet thick, it is improbable that they can penetrate the heavy pack frequented by the Ross seals. Indeed, in the Arctic regions, Eskimos assert that killers will not pursue narwhals and white whales into narrow leads in the ice-pack because of the sensitivity of their 6-foot high dorsal fins, and they are certainly reluctant to enter shallow waters in which they cannot submerge their fins.

Nor, in the Antarctic, are they likely to prey much on the Weddell seals among the thick bay-ice, though it is true that the latter have often been found in the pack-ice of the Palmer Peninsula. Normally, however, these seals do not venture far out to sea because they can hunt for fish beneath the land-fast ice and come up for air at cracks and holes as much as a quarter of a mile in from the seaward edge of the shelves, which slope down landwards to water-level. Possibly the killers are most dangerous to them during the breeding season, and Herbert Ponting (photographer to Captain Scott's polar expedition) saw a Weddell cow plunge off the ice-shelf three times into a pack of killers, in efforts to rescue her pup swimming in the water below; but though she dived beneath it and attempted to hoist it on to the ice with her back, she sacrificed her own life in vain.

It is ironical that although packs of killers range the world's oceans in extraordinary numbers from the Greenland Sea to the Ross Sea, and are often to be seen at the sur-

Weddell seals. Mother seal with pup

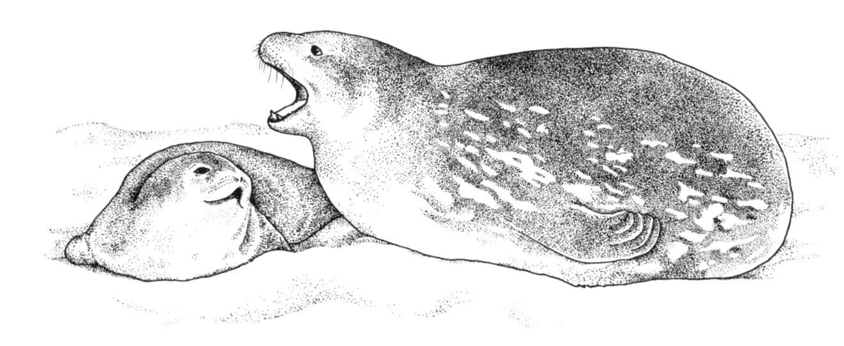

face, we know virtually nothing about their life-history because, unlike baleen and sperm whales, they have no commercial value. Men in general and zoologists in particular will have to alter their attitude radically towards all the cetaceans, for we now know that these animals' brain capacity is equal or superior to our own. One cannot even begin to imagine the mental world of a sperm whale possessing a brain six times as large as a man's; yet the terrible fact is that we continue to butcher sperm whales, and are in some danger of exterminating bottle-nosed dolphins, because of the worldwide craze for dolphinariums, in which these super-animals can perform circus tricks, though their mortality rate during and after capture is high.

There is, however, another powerful predator which could be responsible for inflicting the crab-eater seal wounds with its fearsome battery of pointed teeth in jaws hinged almost as far back as its eyes—the leopard seal. Although the males of this carnivorous species of seal rarely exceed 10 feet in length, females may reach 12 and possibly 14 feet and weigh upwards of 1,000 pounds. The pack-ice is their main habitat, though they are often encountered in the region of the Polar Front and occasionally in tropical seas; in the pack they hunt fish and squid, and also apparently take large quantities of krill, seize an occasional unwary shag, giant petrel or penguin from a floe, and prey on other species of seals. Although they are capable of killing adult crab-eaters on the ice despite the latter's agility, they no doubt prey mainly on younger seals. These include the gigantic juvenile elephant seals, which the leopards perhaps regard as objects on which to exercise their hunting skill rather than as victims, for they have been observed apparently playing with them, albeit imprinting their tough hides with deep teeth-marks. That leopard seals ever deliberately attack humans, as they are popularly reputed to do, is as doubtful as it is of killer whales. They are certainly curious about man's identity, rearing high out of the water to peer at a man standing on a floe or at the edge of an ice-shelf, just as a pack of killers will thrust up their fearsome heads 6 or 8 feet above a floe for the same purpose. Tap gently and regularly with a rowlock on the gunwale or thwart of your boat, and it will not be long before a leopard swims alongside and looks up into your face.

Leopard seals are solitary hunters, sparsely but widely distributed throughout the Antarctic, with a total population of perhaps 100,000, perhaps twice that number. When they haul out to sleep on the ice, each leopard does so on its own floe; when they pup in the early summer far out in the pack, it is in solitude. When during the winter months of July and August some of them leave the pack in order to haul out on island beaches—though never as far up from the tide-line as the tussock-grass or tussac—half-a-dozen may occupy one beach, but they remain well apart; and there will be no more than a score or two on all an island's beaches. Only at Heard Island, more than 1,000 miles from the nearest coast of Antarctica, is there an exceptional winter concentration of as many as 700 leopards. For nine months of the year there are leopards patroling those coasts of Antarctica and the islands which lie adjacent to penguins rookeries, but again each leopard (or sometimes pair of leopards) hunts along its own stretch of coast. Of their relations with those unique birds, the penguins, we shall have more to say in Chapter Nine.

Some zoologists minimize the hostility of the Antarctic environment, on the grounds that only a minority of its inhabitants experience the full severity of its winter climate.

True, the majority of Antarctic birds do not experience extreme cold nor the most prolonged blizzards, because most of them nest in coastal regions where temperatures during their breeding season may be no lower than those they may experience while wintering in the northern hemisphere; though they will have to contend with storms of hurricane force, which strike the islands and the coasts of Antarctica in any month of the year. It is also true that the majority emigrate after the breeding season, about the time that the sea-ice begins to close in on the land, though many travel no further than the pack-ice. But, on the other hand, some millions of penguins, petrels, gulls and albatrosses (many of them nestlings or juveniles), and even a few hundred passerines and ducks, do in fact weather the terrible Antarctic winter, and continue to do so winter after winter, surviving temperatures as low as −60 or −70 degrees F and blizzards of ten days duration. The male emperor penguins, incubating on the ice, must survive such conditions without food for several months. King penguins must provide their young with fish or squid at intervals throughout the winter; so too must the wandering albatrosses, whose nestlings—protected from the blizzards by a mail of feathers growing beneath their down—sit for nine months on their drum-nests of trodden mud and tussac roots (2 or 3 feet high and 3 feet in diameter) sited on headlands where the winds prevent the snow from lying to any depth.

It is also true that while the pack-ice is only less exposed as a habitat than the high plateaus of Antarctica, air temperatures over the pack seldom fall below 14 degrees F, and the crab-eater and Ross seals can find shelter among the pressure ridges and hummocks or, in low temperatures or blizzards, retreat to the comparatively warm waters of the sea. But the

Weddell seals, inhabitating the coastal regions of Antarctica and also South Georgia and possibly Heard Island where very much lower temperatures prevail than in the pack, can only survive the winters by remaining under the ice for days or weeks at a time. When rising to breathe in very cold weather they expose no more than a couple of inches of their muzzles above the water of their ice-holes, and do not slither out on to the ice until the temperature rises to about 8 degrees F on calm days; whereas during the summer months thousands lie out on the ice, basking in the sun in such favored localities as McMurdo Sound below the Ross Ice-shelf. During the winter, indeed, their presence may only be revealed by the prolonged snorts and vibrant trills sounding from the depths of their snow-covered breathing holes or beneath the thin sheets of ice which glaze the tide-cracks. Weddell seals are, as Scott expressed it, great musicians, especially during the breeding season, and able to produce any note from the plaintive piping whistle of a bullfinch to the prolonged deep moaning characteristic of most seals, interspersed with gruntings and gurglings. Tide-cracks are formed by the moving up and down of the sea-ice as the tides rise and fall at its juncture with the land-fast ice, and two or three cracks may run parallel where the seaward slope is only gradual. They are of considerable importance to the seals as natural exits and breathing holes.

The Weddell seals are able to winter under the ice as far south as the Ross Sea, within 10° of the South Pole and in waters whose temperature would be fatal to a man in a matter of minutes, because they can conserve their body heat in a form of vacuum. This is achieved partly by means of a layer of insulating blubber and partly by their ability to reduce the flow of blood to the surface tissues, where body heat is most rapidly lost, by shutting off most of the small blood vessels in the skin from the main circulation, leaving open only sufficient to nourish the skin and prevent it from freezing. They are also able to keep open breathing holes in the thick bay-ice by continually gnawing and sawing at it with their teeth. Herbert Ponting was able to demonstrate with his films sixty years ago that they do not actually *bite* the ice, as they are still often reported to do but, opening their jaws wide and pivoting on their shoulders, they swing their heads from side to side and *saw* or *scrape* the ice away with their incisors and canines. And he observed that even pups a few weeks old would imitate their mothers' sawing technique. Moreover if, when the ice reaches its maximum thickness of 10 feet or more, a seal is unable to project itself far enough out of its hole to rest its flippers on the rim, and thence work its way on to the ice with considerable floundering, it uses its teeth in the manner described to scoop out an inclined trough equal to its own width up which, with a tremendous expenditure of energy it drags itself laboriously, and immediately dries off by rolling over and over in the snow. It has been suggested that the Weddell seal's technique with ice has been gained at the expense of a reduction in its life span to a maximum of perhaps twenty years, whereas crab-eaters and leopard seals are known to live to upwards of thirty years. Since ice is very hard at low temperatures, the constant chiseling wears down the seal's teeth to such an extent that it becomes impossible for an individual to maintain a breathing hole, with the result that it must ultimately suffocate if it is unable to locate another seal's breathing hole or a tide-crack through which it can rise for air. But this seems a doubtful supposition because, in the first place, breathing holes are maintained and shared by as many as fifty of its companions; and, in the second place, the ringed seals of the Arctic, which are masters of the technique, are known to live for upwards of forty-five years.

No doubt there are hazards associated with a submarine existence under solid ice, several feet thick, extending for miles, particularly in localities where there are no tide-cracks. Winter must be a season of continual enforced activity—feeding, ice-sawing, visiting breathing holes—without sleep, except possibly when a seal is fortunate enough to locate a layer of air, trapped beneath a peculiar formation of ice, in which it can sleep beneath the surface. Certainly seals are continually on the move under the ice throughout the winter, and quite often when the members of an expedition have cut a new hole in the ice, a seal pops up only a few minutes later. A number of observers have noted Weddell seals arriving at their breathing holes, which may be as much as half a mile apart, in an exhausted condition, as if they had delayed too long to surface for air. Herbert Ponting has described, in *The Great White South*, how:

> Sometimes a seal would arrive at a hole in such distress that the vehemence of the expiration of its too-long pent-up breath was almost like the blowing of a whale; and the rapid breathing that followed showed plainly that it had remained submerged almost to the point of exhaustion. . . . It was a wierd experience to stand beside one of their blow-holes on a calm, moonlight winter night. The thin film of ice over the water indicated that the breathing-place was in frequent use. Suddenly, a tremor would ripple through the film; then a seal's head would break it and shoot out of the hole, all glistening in the moonlight, and, loudly snorting, its owner would draw in long draughts of the keen, biting air . . . as, furiously panting, it gazed at me with its great soft eyes in blank amazement.

> I have often stood ten or fifteen minutes beside a hole, waiting for a seal's appearance. Once in the winter. . . . I proceeded to a favorite hole and waited. Before I had been there five minutes, a seal emerged, its beautiful head all blazing with phosphorescence and, as the water trickled down its neck, for a few moments it seemed to be covered with little streaks of flame, whilst a ghostly glimmer shone in the hole. After half-a-dozen healthy blows, it shut its dilated nostrils, almost with a snap, ducked quickly below the surface, and disappeared.

There are also hazards associated with sleeping out on the ice, and a Weddell seal seldom does so at a distance of more than a few yards from the nearest water, whether this be tide-crack or breathing hole. Nevertheless, if there are no cracks in the vicinity, and the seals are using their breathing holes as exits and escape-hatches, there is always the possibility that these may freeze over while they are lying out, for the water in the holes can freeze to a thickness of 18 inches in twenty-four hours. In these circumstances a seal may be frozen to death or, as has actually been observed as late in the Antarctic spring as the second week in October, frozen by head, neck and shoulders within a hole.

Climatic factors must also be responsible for a considerable mortality among the pups, which are born from late August to late October in loosely assembled rookeries. These may be sited near islands where tide-cracks afford easy access to the water, or on the land-fast ice where, though perhaps as far as 20 or 25 miles from the sea, there are holes and cracks at water level. The pups (occasionally twins) are suckled for six or seven weeks, during which period they increase in weight from some 60 pounds to 250 pounds, while their mothers are losing about 300 pounds of their own weight. When only a week or ten days old, and possibly before they have begun their thirty-day molt, the pups are able to bathe in the sea-water ponds that collect in hollows among the 20-foot pressure-ridges; or may even be encouraged by their mothers to plunge off the ice into the sea. Others, however, barely move from the spot on which they were born and Sir Douglas Mawson, who was a member of the early Scott and Shackleton expedition, gave an account in *The Home of the Blizzard*, of what must be the fate of many pups. In October a Weddell cow gave birth on a floe. In the weeks that followed there was a succession of 70-mile per hour gales with temperatures below zero fahrenheit, and the pup ultimately became so weakened that it was unable to extricate itself from the hole its body had thawed in the soft sludgy ice. Another hazard faces those pups born on coastal ice, for if this breaks up at an unseasonably early date, 50 per cent of them may perish.

3: The Fount of Life

Virtually all life, of whatever kind, on Antarctica, on the islands, or on the pack-ice is directly or indirectly, wholly or partially dependent for existence on food obtained from the sea. Indeed, it is possible that those primitive lichens in the ice-free valleys and in the mountains hundreds of miles inland from the coast of Antarctica, survive in those polar deserts (and provide a habitat for a few species of minute insects) only because they are nourished by nitrogenous fertilizers borne on the winds from the guano deposits at the coastal bird colonies. The marine food of birds and seals is itself dependent upon the ability of microscopic plant organisms (phytoplankton) to carry out the process of photosynthesis, in which the energy obtained from sunlight is utilized to convert carbon-dioxide, water and mineral salts into carbohydrates, proteins and fats.

In polar seas the duration of photosynthesis is necessarily restricted, since it cannot begin until the break-up of the ice in the early summer permits sunlight to penetrate the water. Favorable conditions may indeed last for no longer than a month, though in areas of converging currents and at the edges of glaciers and stranded ice-bergs, where upwelling bottom water rises with additional supplies of fertilizing salts, they may be prolonged by a further month or two until high summer—but no later. Nevertheless, despite this short season of burgeoning, none of the world's oceans is richer in plankton, especially diatom plants, because of the constant replenishment of nutrient salts, upwelling around the coasts of Antarctica and flowing in surface currents outwards to the Polar Front. The concentration of diatoms may be fifty times greater than in tropical seas, and their billions provide food for the most important animal in the Antarctic, that diminutive claw-less "lobster" less than 3 inches long, the krill *Euphausia superba,* which captures the minute diatoms with the feathery appendages to its fore-parts, while propeling itself with its five pairs of paddles. The distribution of the krill is erratic, but a single concentration averaging one to every cubic inch may extend for 150 miles and include swarms varying in size from a few square yards to half an acre.

Although a monograph of several hundred pages, based on years of field and laboratory research, has been compiled on the krill, there are still many gaps in our knowledge of their life history. But it seems probable that their distribution is determined by the ocean currents, and that they pass their developmental stage during the winter while drifting south in a deep current from the edge of the pack-ice in the region of the Polar Front, the northern limit of their range. By the following spring, when near the coast of Antarctica, they are at an adolescent stage, and thereafter travel north again during the summer with surface currents to the Front. There they reproduce and continue their life-cycle, which extends to upwards of three years. Although supplies of diatoms are almost non-existent during the winter months, the krill are able to survive under the ice, and are to be found among the ice-crystals in the breathing holes of the Weddell seals. Perhaps they subsist on detritus on the sea bottom, or survive without food.

Wandering albatross—male on nest with wings outstretched, female at right

It is the krill that support the millions of sea-birds. Where the krill are most concentrated, sometimes near or actually on the surface—when their reddish-brown swarms follow one upon another like cloud shadows chasing across the land—there will be assembled the greatest numbers of petrels and albatrosses, and also the densest shoals of fish and the largest aggregations of seals and whales; while on adjacent coasts, if the terrain is suitable, will be the rookeries of penguins. There are seven species of penguins in the Antarctic with a total population of several millions. All consume enormous quantities of krill, and the smaller species probably could not rear their young without it. Adélie penguins indeed, are reported to gorge on krill to repletion and then, in the approved tradition of gourmands, vomit and begin again. Krill is also the almost exclusive food of both adults and young of the tens of millions of petrels—the whale birds, prions, Cape pigeons, shoemakers and the remainder of the twenty-four species breeding in the Antarctic. And just as the teeth of crab-eater seals are specially adapted to the retention of these small crustaceans, so on either side of a petrel's upper mandible are comb-like serrations composed of small horny plates of the same substance as its beak. Swinging, turning, banking with concerted movements, a flock of petrels drop suddenly to the sea and almost creep along the water, with heads stretched forward and bodies resting lightly on the surface, but with wings spread above it and pattering feet providing the motive power. Then, thrusting their heads under, they scurry swiftly forward in droves, scooping up the krill with their broad bills, and occasionally plunging two or three feet below the surface in pursuit of small fish. Water and krill are scooped into the deep, expansive pouches of bare skin below their chins. When full, the pouches are contracted, forcing out the water, while the krill are retained by the serrated plates.

Krill is the staple food not only of the crab-eater seals, but also of the southern fur seals, which are now re-establishing colonies on South Georgia and some of the other Antarctic islands after their kind had been almost exterminated by sealers and whalers. The immigration into the Antarctic in October and November of the great baleen whales—blue, fin, humpback, the few remaining right-whales, and the smaller sei and lesser rorquals—is timed to coincide with the summer proliferation of the krill, which multiply with the sudden explosion of diatoms during long days of sunshine. On the krill the whales wean their calves, born several months earlier in warm tropic seas. Do they locate the swarms of krill fortuitously, or are they able to detect the high-frequency vibrations, emitted by the myriads of individual krill, by means of the waxen plugs which, superimposed on their ear-drums, act as sound-conductors?

Nor is it only the higher forms of life that have taken advantage of the Antarctic's superabundance of plankton. Sponges, which one normally associates with warm seas, are actually more abundant than in the tropics. Equipped with protective needle-sharp spicules, they too feed on diatoms. (The bottom fauna also includes Gorgonacea corals, anemones, sea-urchins, bristle-stars and starfish, sea-cucumbers, clams, mussels, limpets and marine worms, 3 feet and even 6 feet long.) Shoals of fish, notably the blue whiting, also gorge on the krill; but though their shoals may include vast numbers of individuals, they are restricted, so far as is at present known, to about 150 species, of which perhaps only three are exclusively Antarctic, or less than 1 per cent of the total known species of marine fish. No doubt many more remain to be discovered, for the Antarctic deeps are virtually unexplored. One of the most numerous, the bull-nosed South Georgia "cod" (*Notothenia rossi*) was, for example, known to science only from one stuffed specimen in a museum at the time of the Royal Geographical Society's *Discovery II* voyage in the early 1930s, though the Norwegian whalermen had long been familiar with it, comparing it to the torsk of their northern seas. About half of the 150 species are these strange cod-like nototheniids. Predominantly Antarctic in distribution, they include the dragon-fish, the plunder-fish, and the ice-fish and the crocodile-fish with enormous heads and cavernous mouths, small bodies tapering to the tail, and large wing-like fins.

No polar fish live permanently in surface waters; but whereas there are no resident populations of fish in the mid-water zone of the Arctic, more than twenty species are known to inhabit this zone in the Antarctic, possibly because the constant upwellings carry up not only nutrient salts for their planktonic food but also slightly warmer waters. Most of the nototheniids are believed to live in the mid-waters or near the bottom; but they must often shoal up into surface waters. Sir Douglas Mawson, for example, caught more than fifty with a handline in a few feet of water off Cape Denison in October; and ice-fish and "cod" certainly feed extensively on krill in the upper waters. The *Discovery II*, when in seas swarming with krill and teeming with thousands of feeding birds, encountered "cod" so closely packed to within some six feet below the surface while gorging on the krill, that their shoals resembled a solid bottom of huge pebbles under the ship's keel. Further evidence on this point is provided by albatrosses which, like gulls, cannot dive but can only dip underwater. Yet parent wandering albatrosses disgorge to their young ones not only squid but also nototheniids up to eighteen inches long.

The suggestion that Antarctic fish do not commonly rise into surface waters because of the possibility of being frozen to death if their bodies came into contact with ice-floes, would not appear to be tenable. The young of some nototheniids actually attach themselves to the smooth sides of icebergs—which harbor diatoms and amphipods—by clinging to them with outstretched ventral fins, while others rest on platelets of ice which float beneath the ceiling of solid ice, and also take refuge in holes and tunnels in these when pursued by Weddell seals. Moreover, the body fluids of nototheniids are reported to contain dissolved salts which enable them to live in water with a temperature well below the freezing-point of blood. Some nototheniids are also partially bloodless. The circulatory blood of the ice-fish, for example, is colorless and semi-transparent, hence the unique creamy whiteness of their gills. Possibly these bloodless fish absorb some oxygen through the skin, as eels do, for in normal fish the red pigment of haemoglobin in the blood is responsible for carrying oxygen from the gills to the body cells; but none of these phenomena are fully understood.

The winter freeze-up certainly does not inhibit the activities of Antarctic fish in any way, for, unlike fish in temperate seas, they continue feeding throughout the winter. Evidence of this was provided by the American biologist Norman B. Marshall who, fishing during the six winter months at depths of from 30 to 120 feet through holes cut in the sea-ice off north-east Palmer Peninsula, caught more than 1,000 nototheniids, many of which had been feeding on fresh krill. Not only was the ice 5 feet thick, but it was also overlaid by snow, and these fish were living in conditions of total darkness, except for a minimum of light filtering through tide-cracks. They must therefore have located the krill by the sensory perceptions of their lateral-lines—those fluid-filled pits in the skin which detect any vibrations in the water, and transmit these via canals to the central nervous system.

The largest of the nototheniids is *N. rossi*. Though few have been caught exceeding 3 feet in length they are known to reach weights of 150 pounds, and are the main prey of the Weddell seals. On one occasion, when looking down through a seal's hole into about eighteen feet of water, Herbert Ponting saw a shoal of these fish lying motionless some three feet above the bottom and was puzzled by the fact that, though capable of moving with lightning rapidity, they made no apparent attempt to escape from a seal which, undulating almost imperceptibly with the propulsion of its tail flippers, snapped up several of them. How do these seals, feeding not only on fish, squid and krill but also on such bottom edibles as sea-cucumbers and large sea-slugs, detect these in the dark waters under winter ice? They are known to be able to dive to depths of 2,000 feet and to remain beneath the ice for periods of 45 and even 60 minutes; and to submerge long enough and dive deep enough to travel upwards of three miles under ice 560 feet thick. Indeed, members of a Graham Land expedition came upon an adult male, in good condition though with an empty stomach, lying near a small hole at the bottom of a 40 to 50-foot deep rift in the ice about a dozen miles from the seaward edge of the ice-shelf. No other crack could be seen on the ice; but if the seal had reached the rift by swimming under the ice, it would have had to dive to a depth of more than 300 feet at the seaward edge of the shelf.

It is well established that seals in general possess acute hearing; and it is probable that their eyes are sensitive to very low light intensities, enabling them not only to take advantage of what little light there may be in the depths, but more particularly of the bioluminescence with which so many fish, cephalopods and euphausians are equipped. The northern race of fur-seals for example, when on migration, feed extensively on the illuminated squid and lantern-fish which rise into surface waters at night; while sea-lions will shoot up to capture flying-fish milling around a night-light. Since it is very much easier to see one's prey when looking up at it from below than when looking down on it from above (and also to approach it from the "blind side") a hunting sea-lion swims on its back. Even when an experimental prey has been lowered to a depth of 240 feet the seal has still approached it by first descending below it and then shooting up at it. So far so good, but when blind seals are caught they are invariably in good condition, and have therefore been able to obtain adequate food without the aid of vision. Moreover, there can be no natural light at any depth beneath the heavy Antarctic ice, only bioluminescence. An American biologist, Carleton Ray, has described in *Animals*, how even in brilliant summer sunshine it is somber and overcast beneath the ice, whose irregular undersurface, stained light brown by algae or diatoms, resembles tight cloud cover. The seals' holes, however, shine like lighted windows and the

Fur seal bull

water in the immediate vicinity of the observer is of wonderful clarity and tinged pale green by the sunlight filtering through the ice, though shading at a distance into dull metallic blue and ultimately blackness. But even in the early summer maximum visibility is no more than 200 feet, and by midsummer has been reduced by the bloom of plankton to 30 or 40 feet. Although Ray and his team experimented with hydrophones and an observation chamber beneath the ice for two months in the fore-summer of 1964 without obtaining proof that Weddell seals are equipped with sonar, an affirmative result was not really to be expected in the prevailing conditions of maximum visibility underwater at the lightest season of the year. The seals' loud chirps and trills, detectable at a distance of 5 miles under the ice are, however, typical sonic sounds; and sonar or echo-location would seem to be the only solution to the problem of how seals in general locate their prey at any considerable depth beneath the icc and during the darkness of the polar winter. But this is a problem to which we shall return when considering the habits of Arctic seals.

At this stage we can begin to comprehend the pattern of life in the Antarctic. It begins with the direct food-chain of diatom→ krill→ fish, bird, seal or whale, but indirectly, though no less vitally dependent on the diatom/krill food base, are the predators. Skuas rob the penguins and petrels of their catches of krill and fish, and also kill them. Leopard seals prey on fish, penguins and other seals. Killer whales prey on seals, including the leopard seals, and possibly on the smaller krill-feeding whales. Only those insects living among the lichens of the hinterland, and the rotifers and "water-bears" in the frozen lakes, are perhaps totally independent of the sea, and have no place in the food-chain.

4: Life on the Antarctic Islands

The influence of the mass of cold Antarctic water on the flora and fauna of those islands lying to the south of the Polar Front is dramatically illustrated by the contrast between glaciated South Georgia and the Falkland Islands, which are less than three degrees of latitude apart but on either side of the Front. The Falklands have a relatively mild though excessively windy climate with a monthly mean temperature above freezing, and are free of sea-ice throughout the year. Their flora includes upward of 150 species of flowering plants, insects are abundant, and approximately two-thirds of the 65 species of breeding birds obtain their food from the land. By contrast the mean monthly winter temperature on South Georgia is below freezing; there are less than 20 flowering plants on the island and these are predominantly tussock-grasses and small recumbent shrubs, for there are no full-sized shrubs on any of the islands in the region of the Polar Front. Insects are scarce, comprising only 5 species of beetles, 4 diptera and 5 springtails; and only 2 of the 25 breeding species of birds feed off the land.

Although summer temperatures throughout the Antarctic are consistently below the minimum necessary for the growth of most plants, this factor must not be overstressed; there are other factors to be taken into account. In the Arctic, for example, there is vastly more snow-free and ice-free ground available for its flora of several hundred seed-bearing plants, with a consequently greater range of habitats; while the continuity of land there provides routes for the immigration of plants able to survive under polar conditions. In the Antarctic, on the other hand, immense stretches of ice-cooled waters separate polar lands from the nearest reservoirs of plants.

No less influential is the pack-ice. A submarine ridge, the Scotia Arc, linking the southern tip of South America with the northern tip of Palmer Peninsula, projects above the sea in a series of islands. Those lying within the compass of the winter pack-ice—the more southerly of the South Sandwich Islands, the South Orkneys and the South Shetlands—experience much longer and more severe winters than does South Georgia, and their mean annual temperature is no higher than that of those regions of Antarctica ten degrees further south. By contrast, strong winds prevent the pack from reaching and congealing South Georgia, there is no regular formation of land-fast ice, and the heavy winter snowfall disperses in October or November. The snow melt is associated with a rapid thaw on the coastal plains, laying bare the ground for such breeding birds as gentoo and macaroni penguins, Antarctic terns, brown skuas and Dominican gulls (the southern black-backs); while banks of moss, tussac and shale soften sufficiently for petrels to excavate their burrows. The further an island from the continent of Antarctica the less severe its climate. South Georgia, Kerguelen and Macquarie all "enjoy" temperatures above freezing for at least half the year, with a summer maximum of 64 degrees F on South Georgia, though even in January (the Antarctic midsummer) the latter island can be bedevilled by eighteen days of snow and rain and fifteen days of gales.

The absence of pack-ice from these islands outside the

Scotia Arc also ensures a permanent reservoir of food in the luxuriant beds of that brown alga, the kelp. Anchored by powerful branching roots to large stones or rocks on the sea-bed at depths of up to 80 feet the kelp rises to the surface on long slender stems which sprout a profusion of jagged-edged fronds from 6 to 24 inches long, and continues to grow along the surface to lengths of 100 feet or more. The dense carpet of interlaced fronds and stems provides shelter and food for countless millions of molluscs, crustaceans and other small invertebrates, and is the constant feeding ground of all manner of birds.

The fact that South Georgia seas remain open all the year round makes it possible for such birds as king penguins and wandering albatrosses which are, as we have seen, obliged to provide food for their young periodically throughout the winter months, to breed on that island, but not on the more southerly islands set in the frozen sea of the Scotia Arc. However, small numbers of king penguins have been able to colonize the more northerly of the eleven small islands which make up the 193-miles-long, bow-shaped chain of the South Sandwich Archipelago where the influence of the pack-ice is comparatively slight, but not the South Orkneys and Shetlands which are engulfed in ice for several months and virtually linked to Antarctica by an almost continuous sheet of pack-ice. The active volcanoes on some of the Sandwich Islands melt the snow and allow immense numbers of chinstrap and gentoo penguins to establish summer rookeries amid luxuriant "swards" of the green algae *Prasiola crispa*, which thrives on exceptionally heavy concentrations of guano at sea-bird colonies in both southern and northern hemispheres. On the volcanic island of Zavodorsky, for example, there may be as many as 10 million chinstraps. Thus,

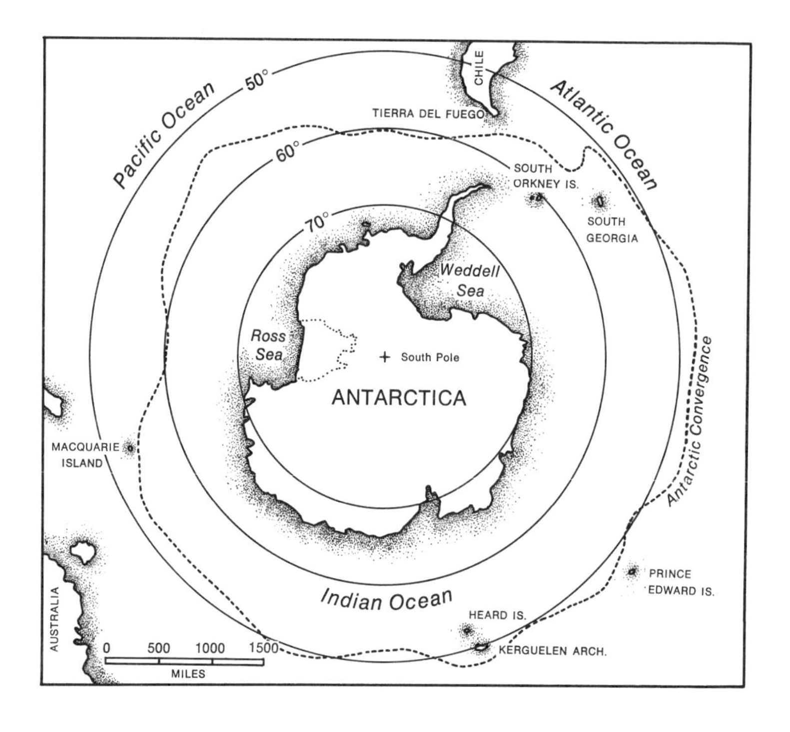

although the winters on some of the islands of the Scotia Arc may be as long and hard as on Antarctica itself, the vastly greater areas of snow-free ground in the summer result in these, rather than the continent, being the metropolis of the huge Antarctic population of breeding birds.

If the extent of the pack-ice determines in a general way the flora and fauna of the Antarctic islands, it is the ubiquitous tussac, reaching its southerly limit in the less severe climate of South Georgia, which is of predominant regional importance for, on the islands of the Scotia Arc, the only vegetation comprises thick banks of peaty moss. Waist-high, even shoulder-high, the green "meadows" of tussac cover most of the low ground on South Georgia, and also on those islands in the region of the Polar Front, and extend 1,000 feet or so up favorable seaward slopes. Each tussac plant grows by sprouting new stems from the stool formed by the decay of older stems, and in the course of years it becomes a huge tuft with a centre of spiked stems surrounded by the drooping outer blades. Melt-water, penguins tramping to and from their rookeries and the wallowing of elephant seals up to 5 tons in weight ultimately excavate channels below the "canopy," and the tussacs are bisected into immense pedestals of earth weighing a ton or so and bound together by their fibrous roots, above which wave the overlapping crowns of coarse green blades. One might as well attempt to force a way through an Amazonian jungle protested the British zoologist Harrison Matthews.

The tussac prohibits the growth of other plants, though beyond its boundaries are bedstraw and dwarf buttercups whose scalloped-edged leaves and starry gold-petalled flowers colonize newly abandoned elephant seal wallows, while on South Georgia mats of the short leafless stems of two species of burnet thrust up hundreds of minute pom-poms through dwindling snow-beds, especially around the bases of stony screes away from the salt spray. Their dark-red or purple-pink flowers add a faint tinge of color to the monotones of the mountainsides, otherwise brightened only by the reds and yellows of some 200 varieties of mosses and lichens. Here and there on lower mountain slopes, and also among the rocks on the shore, three or four species of fern are to be found, including the extraordinary brittel fern (*Cystoperis fragilis*) whose other known habitats include not only the Falkland Islands and Chile but Mount Kilimanjaro, Abyssinia, and Spitzbergen.

It is only of incidental importance, though none the less

significant, that it is the tussac that has made it possible for stocks of reindeer, imported to South Georgia in 1911 and 1912, to thrive on the rolling hills and, with winter shelter in the deep valleys, to increase in numbers to several hundred. Their survival indicates that supplies of food on the more northerly of the Antarctic islands are sufficient, despite the heavy snow covering during the winter, to support indigenous mammals, and that the absence of the latter must be due to some other factor—presumably the absence of any land-bridges across the Antarctic Ocean. For instance the ubiquitous brown rat has been able to survive the winters on South Georgia in the shelter of the tussac since about 1800, and far from being dependent for food supplies on man in the form of refuse from the whaling stations, apparently subsists mainly on the stems of the tussac and on petrels in their nesting burrows. Horses have also been able to survive on South Georgia, for some imported by a German expedition in the early years of this century throve and bred in a feral state for at least seven years on magnificent pastures of shoulder-high tussac. Although sheep and rabbits are reported not to be able to withstand the winter blizzards on South Georgia, a small number of the latter have apparently survived on one islet. Predation by skuas has perhaps been as decisive a factor as the climate in the failure of the rabbits, though it is true that they are still abundant on Macquerie and Kerguelen where skuas also breed. But it may be their grazing habits rather than the climate which has prevented flocks of sheep from becoming established on South Georgia. Where stock are pastured on the Falkland Islands, for example, the tussac disappears because both sheep and cattle tear up its succulent nutty-flavored stems; and a small flock of sheep on a South Georgia islet did actually clear the tussac to such an extent that it took several years to regenerate.

Of more direct concern is the fact that, as more than one ornithologist has noted, of ten species of birds breeding on South Georgia, but not in the South Orkneys, at least eight are able to do so because the tussac provides them with food, shelter and nesting sites. Thus the shoemakers and diving petrels which nest in holes excavated in tussac banks, the gray-backed storm petrels which commonly nest among the roots of the tussac, and the albatrosses which favor tussac-covered hillsides and cliffs, are all absent from the southern islands of the Scotia Arc.

It may well be the tussac that has preserved the South Georgia pipits, which closely resemble the rock pipits and water pipits of North Atlantic shores, and like them parachute down from short trilling song-flights. There is no evidence that any predators are interested in such small birds. The pipits indeed almost run over the toes of skuas on the beaches, and some build their nests only a few feet from those of the giant petrels which scavenge on anything, alive or dead. But it is difficult to believe that they could have survived in the Antarctic without the shelter provided by the tussac, whose fine roots, together with portions of leaf, supply them with material for their nests which are built among the tussac or in rock crevices on low cliffs. The tussac also furnishes the pipits with a certain amount of vegetable food, though no doubt these birds feed primarily on the insects engendered in the wallows of the elephant seals and in the rotting beds of storm-wracked kelp, on the floating beds of which they are often to be seen standing feather-deep in the water, catching amphipods for their nestlings. In the winter, however, when the island is blanketed by six or ten feet of snow, only the beaches can provide them with food, and they follow the ebbing tide closely, picking up one morsel after another before the wet sand becomes glazed with ice.

Those other curiosities on South Georgia, and also on Kerguelen—the teal or pintail ducks—also frequent the tussac, guzzling in its damper parts and in the elephant seals' wallows, and swimming in the small melt-water pools that abound in the summer, when the lower ground is comparatively free of snow. In the winter, however, they, like the pipits, can only find food among the rocks and freezing seaweed between the tide-lines on sheltered beaches. Though formerly commonly distributed throughout the coastal regions of South Georgia, despite predation by skuas, their numbers have been decimated by whalers and sealers.

It is the tussac, providing admirable resting, wallowing and molting retreats, that attract the monstrous elephant seals to the islands and particularly to South Georgia, where half the Antarctic population of some 600,000 have their headquarters during the breeding season in the early summer, and also when molting in the autumn after a short period at sea. Feeding mainly on squid and other cephalopods, for which they may dive to depths of 2,000 feet during 10-minute submergences, the elephant seals are widely distributed through the Antarctic, following the pack-ice southwards as it recedes in the summer, and perhaps ranging along its northern edge in the winter, though also wandering far up into the roaring forties. But only an occasional individual strays to Antarctica, where there are no beaches on which a herd could haul out, and no tussac in which to slumber. It is the pack-ice that prevents the elephant seals from colonizing the islands of the Scotia Arc, though some bulls do in fact maintain breathing holes in the sea-ice for a few weeks during the spring and autumn. A few harems are also established on the sea-ice, but if the latter breaks up during the breeding period more than three-quarters of the pups may perish. Nor is life easy for the pups on South Georgia. If the spring snowfall has been very heavy and lies eight or ten feet deep, the cows may be compelled to bear their pups on the beaches just above the high-water mark, with the result that many of the latter are carried away by storm breakers, or squashed by the bulls lumbering across the beach in pursuit of younger rivals, which are seldom allowed to approach the cows. Even in the rookeries in the tussac large numbers of pups may become trapped in holes from four to six feet deep, because the heat of their bodies—or perhaps the heat of the sun absorbed by their coal-black woolly fur—melts the deep snow. In these holes the pups cannot be suckled, and must starve to death.

5: Astounding Antarctica

An orderly survey of the ecology of Antarctica is upset by the conformation of Palmer Peninsula, the tip of which extends to the same latitude as the South Shetlands, with the result that its mean winter temperature may be as high as 27 degrees F and rise to above freezing-point for at least one month in the summer, with a maximum recording of 53 degrees. Although hurricanes can break up the ice and produce temporary open water in winter as far south as McMurdo Sound in lattitude 77 degrees, the coastline south of the peninsula is normally fringed with thick ice for ten or more months every year, and even under the most favorable conditions snow-free ground is only to be found in narrow strips at the edge of the sea for two or three months in the summer. On the west coast of the peninsula, however, there are small havens where there may be open water, though no kelp jungles, for all but a few days in the coldest months, enabling resident populations of cormorants and giant petrels to feed at sea even at midwinter. So, although 17 of the 43 species of birds breeding in the Antarctic have been able to establish colonies on Antarctica (7 species of petrel, 4 penguins, 2 skuas, Dominican gull, Antarctic tern, blue-eyed cormorant and sheathbill), 4 of these (gentoo penguin, sheathbill, cormorant and brown skua) are restricted to the peninsula. Moreover the dove prion (one of the petrels) has apparently colonized only one station south of the peninsula;

chinstrap penguins only isolated ones among the rookeries of Adélies in north Victoria Land; and only three species (emperor penguin, McCormick's skua and Antarctic petrel) breed exclusively on Antarctica. Since, incidentally, new colonies of penguins, and also petrels, are being discovered on Antarctica every year, it seems too early to agree with the suggestion that the chinstraps are increasing in numbers and extending their range because the 85 or 90 per cent reduction in the number of whales from the early years of this century has resulted in a significant increase in the stocks of krill.

With so little snow-free ground available on Antarctica, every accessible niche and ledge on exposed coasts—where the cliffs and headlands, scoured by the ceaseless winds, are sometimes more black with rock than white with snow—is tenanted by nesting birds, especially in those localities where the sea-ice melts comparatively early in the summer, though remaining intact elsewhere, and opens up small fishing polynias (or pools). Giant petrels nest on open lichenous flats, with smaller species burrowing beneath them; while silver-gray petrels (the southern fulmars) inhabit the steepest rock-faces, exposed to the full blast of the blizzards and in situations which may involve them in having to work for hours at clearing the snow from their nests with their beaks. Moreover if a nest becomes filled with frozen snow the warm egg may melt a hole in this and drop through, with the result that the fulmar is unable to incubate the egg. But given the rare calm midsummer night, the cawing of the fulmars and the sweet bell-like piping of numberless small petrels in their burrows sounds from every headland.

But what life can exist in the frozen desert of Antarctica's hinterland, 96 per cent of which is buried beneath the ice-cap where waves of ice follow waves of ice, as though to deride the fact that men called this a plateau? On those high ice-plains the mean annual temperature never rises above freezing-point. It is −58 degrees F at the 9,000-foot depression of the South Pole and −71 degrees F on the 14,000-foot heights of the Pole of Inaccessibility, while at Vostok, at an altitude of 11,000 feet and some 400 miles in from the coast of Queen Mary Land, a temperature of −127 degrees F has been recorded—the lowest known on Earth. The plateau air is so cold and dry that there is no rain, and the annual precipitation of snow, falling as crystals, is less than 2 inches. On a day of sunshine the sky may be clear overhead, with only a faint haze around the horizon; "yet suddenly, and apparently from nowhere," wrote Scott in *The Voyage of the "Discovery"*, "a small shimmering body floats gently down and rests as lightly as thistledown on the white surface. The six-pointed feathery stars, flat and smooth on either side, rest in all positions, and therefore receive the sun's rays at all angles . . . reflecting in turn each color of the spectrum. . . . The gem-like carpet . . . sparkles with a myriad points of brilliant light, comprehensive of every color the rainbow can show."

Nevertheless, even on the high plateaus it can be almost hot on a calm sunny day. Scott was tragically unlucky in the weather he encountered during his trek to the Pole, for Amundsen, preceding him by only five weeks, was skiing in his underwear and lying outside his sleeping-bag when only four degrees from the Pole. Scott could not know that the route he had chosen up the Beardmore Glacier—the generally accepted gateway to the Pole—lay through the regular blizzard zone on the western half of the Ross Ice-shelf, while, with a dramatic irony matching the conditions,

Amundsen, by sportingly refusing to encroach on Scott's territory, unwittingly selected the best-weather route up the Axel Heiberg glacier. His account of his trek reads like a picnic—which it was not. But then the Norwegians were professional explorers, professional on skis, professional with their dogs; whereas the British were, as ever, heroic amateurs, additionally handicapped psychologically by their constant remorse for the sufferings of their ponies, and by the emotional stress to which they were subjected by each successive butchery.

In contrast to the coastal regions very light winds prevail in the deep interior of Antarctica, with storm gusts rarely exceeding 70 miles per hour, or so meteorologists assert, though one American observer did not record a single hour of calm, even with temperatures below −94 degrees F, during the course of a complete year at the Pole. In ideal conditions snow-free rock surfaces may heat up in the sun to temperatures as high as 83 degrees F and provide habitats, less than 300 miles from the Pole, for some of the few hundred species of lichens which, together with mosses and algae, comprise the only forms of vegetation on Antarctica south of Palmer Peninsula, where a pink and two hair-grasses have been able to establish themselves in loam-like, though worm-less soil. In a few localities, indeed, there is the rare sight of a hillside bright with lichens—orange and red, black and white, yellow and gray—or carpeted with an almost luxuriantly green growth of mosses. On the under surfaces of lichen-encrusted rocks primitive insects have by some miracle, since they are wingless, founded colonies. One-twentieth of an inch long, blue, black, brown or in some instances white, these springtails can survive temperatures as low as −58 degrees F, providing that the humidity is not too low. If temperatures fall below this degree they become dormant at whatever their particular stage of development until a rise in temperature reactivates them. In this way they somewhat resemble the microscopic "water-bears" and bright red rotifers, whose habitat is the algae in ponds and small lakes which are almost permanently frozen. These, the most extraordinarily adaptable forms of life on Antarctica, not only survive being frozen in their lakes for periods of years, but can also survive the ordeal of being removed from the ice, thawed, dried out, and then immersed in water near boiling-point—all within the space of a few hours, during which period they have been subjected to a temperature range exceeding 200 degrees F. It is a comfort to know that if man extinguishes all other life, including himself, on Earth, there will probably still be indestructible "water-bears," whose distribution is global. The German biologist, Vitus Dröscher has described how water-bears have been exposed to a stream of hot air with a temperature of 197·6 degrees F and when subsequently moistened at room temperature were lively within half an hour. They have been fused into glasses lacking oxygen; imprisoned for months at a time in pure hydrogen gas, nitrogen or helium, in carbonic acid, hydrogen sulphide or coal gas, and have revived when returned to water. They have been sealed for a period of twenty months in liquid air at a temperature of about −392 degrees F and then in liquid helium at an "outer space" temperature of −519·8 degrees F, and still they have thawed out as lively as crickets!

Except on the coastal cliffs and in certain "dry" valleys Antarctica's underlying rock is bared only on isolated peaks, or nunataks, protruding above the snow and on the higher mountain ranges, and even the latter have attracted life. As long ago as 1912 a Douglas Mawson party, when camped at a height of 2,450 feet and 76 miles inland from Cape Denison, was astonished by the appearance during a 65-mile per hour gale of two snow petrels which, hovering with outspread wings just above the snow and touching it with their feet from time to time, were able to maintain a stationary position without a movement of their wings, just as they do when picking krill off the ice-floes in strong winds. We now know that both snow petrels and Antarctic petrels actually nest on the crags of mountain ranges as far south as 77°. No doubt they return to them in September or October, when their relatives in coastal colonies arrive from their wintering areas in the pack-ice, though on the west coast of Palmer Peninsula dozens of snow petrels are already flying around the cliffs and alighting on the ledges as early as August. In mid-December, when 200 miles inland from Shackleton Base, members of the Trans-Antarctic Commonwealth Expedition saw hosts of snow petrels wheeling around the 2,000-foot cliffs of Mt. Faraway in the Theron Mountains of Coats Land, and found them nesting in the rocky crags, with skuas in the lower fastnesses—an interesting example of colonization by a parasitic species, for the skuas must presumably have been preying on the petrels. A month later two snow petrels were seen flying around the top of a 7,000-foot peak 230 miles inland from Halley Bay. No doubt they nest there too, as they (and also skuas) are known to do on north-facing cliffs, which catch the sun, in the 6,000-foot Sør-Rondane mountains nearly 200 miles from the coast of Queen Maud Land. Is the competition for nesting sites on the limited stretches of coast available for breeding birds so intense that these petrels must colonize crags at such distances from open water, to which they will have to travel in order to obtain krill for their nestlings? These long flights oblige the parents to take alternate spells of incubation and brooding, while one or the other is away fishing in polynias (open pools) or leads in the ice but, even so, they must face problems in keeping the nestlings supplied with food. Moreover, if the ice is late in breaking up, and the parents are unable to find open water within tolerable distances of the breeding crags, the nestlings must starve. This they may also do if blizzards delay the return of parents or bury those nests in exposed situations and not in the more usual tunnel-like holes up to 6 feet long under slabs of rock. Or is some other factor responsible for the establishment of these inland colonies? Environmental conditions at snow-petrel habitats outside Antarctica—notably on Bouvetøya and the South Orkneys, where there is a large population—are, for example, almost as severe as in the interior of Antarctica.

Fresh water the petrels and skuas will not lack, if they require it; and the skuas of Antarctica are as markedly addicted to bathing in fresh water as are their relatives in the northern hemisphere. There is a surprising amount of summer melt-water in the interior of Antarctica. Scott was surprised to find that in the lower reaches of glaciers, where the air temperature in the valleys was often above 40 degrees F in January, streams were flowing 9 inches deep and every boulder was standing in up to 3 or 4 feet of water. So too, Sir Vivian Fuchs reported in *The Crossing of Antarctica* that a party from the Trans-Antarctic Expedition, when approaching in December a great barrier of crags intersected by steep glaciers in the Theron Mountains, experienced difficulty in crossing a broad stream of melt-water, which was rushing into a small lake that had formed on the surface of the ice. And they were astonished in so high a latitude (near 80° S) at the amount of water pouring in miniature waterfalls off the precipitous rock and ice slopes of Mt. Faraway to feed the river flowing along its base: "As the day advanced and the sun sank," he wrote, "it seemed that an invisible hand slowly shut off the supply of water, reducing the spouts and rushing torrents to mere trickles that gurgled their way beneath screes or dripped slowly from the hanging masses of the tumbling ice-falls."

The nunataks of protuberant rock are most numerous on the coastal edge of the hinterland; and between the nunataks and the coast, in some localities, ice-free cliffs rise to heights of more than 3,000 feet above valleys that were formerly glaciated, but in which the glaciers have withered so that they now contain no banks of snow, no large masses of ice. Their steep bare hillsides are varicolored like those in other parts of Earth, russet-brown, bright-red, dull gray striated here and there with bands of yellow, while the brick-reds of the deep gorges near the summit of the Ferrar Glacier recall the Grand Canyon of the Colorado. For that matter, the overall whiteness of Antarctica seldom looks white, since it embodies shades of many hues, especially cobalt-blue or rose-madder and all gradations of lilac and mauve produced by the mixing of these colors.

The floors of these dry valleys—the "oases" of Antarctica—may be confused jumbles of boulders or undulating stretches of sand traversed by the silver threads of numerous small streams, which flow into shallow lakes, some ice-free, some frozen and gleaming white, while several square miles of one of the eight or nine Ferrar valleys are uniformly patterned with dark polygons, resembling honeycomb with cells 60 feet in diameter. A familiar polar phenomenon, the polygons have been created by the repeated thawing and freezing of waterlogged silt or gravel, which expands on first freezing, contracts when the temperature falls again to sub-zero levels, and then cracks into discs. Snow, filling the cracks, preserves the mosiac when the ground expands again. Although the air temperatures in the valleys are normally near freezing throughout the summer, they can rise as high as 54 degrees F because the rock surfaces absorb more heat than ice or snow; while the lakes, on whose surfaces puddles of summer melt-water form and refreeze, serve as natural heat reservoirs, whose saline bottom waters may be warmed to 77 degrees F at depths of more than 200 feet.

The largest known of these oases is McMurdo on the mainland opposite Ross Island. Through the crumbling brown hills and steep-walled valleys of its 1,500 square miles, the thin sliver of the Onyx river flows briefly in summer. No birds nest in it, though an occasional skua passes through—as skuas probably do over most parts of Antarctica, since they have been recorded within 150 miles of the Pole. On January 2, 1912, for example, one kept alighting ahead of Scott's polar party when they were at 87° 20′ S; and a week later two settled for a while at Amundsen's camp at 84° 30′ on the Ross Ice-shelf, before continuing on their way south, as if intending to traverse the continent from coast to coast. Edward Wilson believed that these visiting skuas were attracted by the scent of blood, because the day after a dog had been slaughtered at a camp above 80° S on the ice-shelf and 170 miles from open water, they were visited by a skua—as they were at the same place on their return journey. This explanation of the skua's presence seems somewhat improbable, though many years later an American party believed that the smell of their cooking attracted skuas to the Bunger Lakes oasis on the Davis Sea coast. Navigation over the white plains of Antarctica probably presents few problems to skuas, for one of the six ringed birds released at the South Pole after an 825-mile plane flight returned to its breeding site within ten days.

But if there is no resident animal life in the McMurdo and other smaller oases they have been, and still are, regularly visited by seals. The bleached skeletons and mummified remains of Weddell and especially, and surprisingly, crab-eater seals, and also those of Adélie penguins, have been discovered not only in the ice-bare oases but on glaciers 35 miles inland and at an altitude of 3,000 feet, and also under several feet of glacier ice at heights of up to 6,000 feet on Mt. Discovery. On his first expedition, Scott chanced upon two cadavers of Weddell seals at a height of almost 5,000 feet and more than 50 miles of rough and steep ascent from the coast; while a party from his second expedition found no fewer than thirteen dead seals in one of the valleys off the Ferrar Glacier, where the moraines were covered with seals' bones. In the 1957–8 season two American scientists discovered no fewer than ninety mummified crab-eater seals at an altitude of between 1,000 and 2,000 feet in the Taylor Dry Valley. Radio-carbon analysis suggests that some of these seals may have died in the oases between 1,600 and 2,000 years ago. This, however, must be considered a provisional estimate because, as a consequence of the constant upwelling of bottom water off the coasts of Antarctica, "old" carbon remains in circulation for long periods, with the result that fish caught today may give a radio-carbon dating of 1,200 years. Determining the age of fish-eating seals by this method is therefore suspect.

Of more immediate interest are the problems of why these seals came to the oases, and by what means they were able to travel such long distances over such difficult country.

Weddell seals progress extremely laboriously over the ice, while crab-eaters, though much more agile, do not commonly haul out on ice-shelves or beaches. One suspects that all have been old or sick beasts, probably bulls, making their painful pilgrimages in order to isolate themselves from their fellows, as so many wounded animals have a strong urge to do. After fighting during the rut, wounded Weddell seals are known to retreat to secluded places as far as 35 miles from the coast, negotiating chaotic jumbles of pressure-ridges and hummocks one would deem impassable to any seal, let alone sick ones. A bull with suppurating sores has been seen more than 20 miles from the sea on the Koettlitz Glacier at the base of Mt. Discovery. And it was on this glacier that towards the end of February members of Scott's last expedition encountered a number of live seals, some sleeping, which had apparently made their way up to it via a freshwater stream flowing alongside and beneath the glacier. If the bulls seek retreat, it is to be found inland and not in the sea with its leopard seals and killer whales.

Although the majority of skeletons have, by all accounts, been those of crab-eater seals, there does not appear to be any record of a living crab-eater being seen farther away from the sea than the coastal ice, and their presence in the oases poses even greater problems than that of the Weddell seals. It has been suggested that in either case they may have wandered inland when there was still ice far up these dry valleys and starved to death, or that they have been forced to retreat from the sea in order to escape exceptional persecution by predators, though the latter solution is hardly applicable to Weddell seals and does not, in any case, explain why they had travelled so far inland. Possibly the curious occurence of several herds of crab-eaters totalling some 3,000 individuals, which in 1955 wintered on the ice of a 50-mile length of the Prince Gustav Channel off the east coast of Palmer Peninsula, has some significance. Ordinarily crab-eaters winter on the pack-ice, and although they are continually on the move within it, occasionally in large herds, this concentration was at least ten times greater than any that would normally be found on coastal ice. It is perhaps explained by the activities of some 120 lesser rorqual whales, 60 killers and a few bottle-nosed whales which, cut off from the open sea by the formation of a 40-mile-wide barrier of sea-ice, had kept open several polynias (or pools) in the channel. Possibly an abnormal concentration of krill at the entrance to the channel when the sea-ice was forming in April could have attracted both seals and rorquals and their concentration in turn have attracted the killers. Alternatively, the several herds of seals—the majority of whom bore the usual scars and lacerations—could have been seeking refuge in the channel from the killers though, remarkably, neither seals nor rorquals were ever seen to be attacked by the killers in the polynias.

The whales had apparently been stranded as early in the autumn as April in a series of pools over a distance of several miles, and on midwinter day the members of a survey expedition from the Falkland Islands Dependencies base at Hope Bay, were astonished when crossing the frozen channel to see a rorqual whale shoot up 7 feet through a hole in the ice. Towering above them, it exhaled with a deep sigh and sank back into the water. Thereafter, the party came upon numbers of holes in the ice, including one area perforated by about twenty holes from 4 to 10 feet in diameter, in one of which four rorquals were blowing, and subsequently five open pools varying in size from a few hundred square yards to some six square miles. Whales were cruising around the leaden black waters of the pools, blowing and diving leisurely, while lying out on the ice surrounds were thousands of silver-gray crab-eater seals and some black Weddell seals. In the smallest pool, which was 100 yards long and 50 yards broad, there was the extraordinary scene of a dozen Adélie penguins, four crab-eaters, a leopard seal and two killers swimming around in apparent harmony. Moreover, when another five killers appeared from under the ice less than six feet from the Adélies, they "porpoised" straight through the penguins and seals, and plunged under the ice on the other side of the pool.

By the second week in August all the smaller pools were frozen over, and the large polynia had been reduced to two pools, each about 350 yards long and 150 yards wide. The rorquals and killers were now concentrated in these, with crab-eater seals on the ice; but though some of the latter were swimming in the pools they were again ignored by the killers. Indeed the latter were apparently on playful rather than predatory terms with the seals; for when a pair of killers accompanied by a 10-foot-long youngster approached a seal which was lying half in and half out of the water, the seal snapped at them and they backed off; and when they persisted in attempting to nip it, the seal finally slipped into the water and brushed past them fearlessly.

With the freeze-up continuing, the two polynias were ultimately restricted to a length of 5 yards apiece, and the members of the expedition were able to pat both rorquals and killers! None of the killers were seen after August and, if they survived, must have reached the ocean by swimming under the 40 miles of ice, surfacing for air at small cracks, none more than a foot wide. If any of the rorquals survived

—and four were blowing in a 10-foot by 15-foot pool on October 21—they did not finally escape until November when the sea-ice broke up. Be that as it may, a score or more of the crab-eaters had traveled over the land-fast ice to a distance of more than 4 miles from the nearest pool and one, in the younger age-group less than nine years old, was 15 miles inland and some 55 miles from open water, and still moving away from the pools when last seen. Now, many of these seals were probably sick, for it was noticed in August that some had aborted their pups and that others were sluggish and lethargic; by the spring, 85 per cent of them in the younger age-group had died and were found to be suffering acutely from both congestion of the lungs and nephritis. This sickness could be ascribed to a highly contagious virus infection, possibly the result of overcrowding or of malnutrition, for there did not appear to be any krill in the pools, though the infection did not spread to the 300 Weddell seals which also wintered in the channel. The first symptoms of the disease may have been the cause of the crab-eaters leaving the pack-ice.

6: Rookeries on Ice and on Land

As late in the autumn as May, downy young snow petrels are still flying around the cliffs of Terre Adélie, and young giant petrels, abandoned by their parents, are attempting to launch their heavy bodies into flight from ice-floes. But if those petrels nesting on the mountain crags of Antarctica must survive blizzards of hurricane strength which blast them with driven snow as dry and granular as fine sand, and must in some instances travel 200 miles or more to the nearest fishing grounds, their problems are relatively insignificant in comparison with those surmounted by the quarter or half million emperor penguins, which breed exclusively around the coasts of Antarctica between 66° and 77° S, and never normally migrate out of Antarctic waters. Soon after the new sea-ice has formed in the early autumn, when Antarctica begins to darken with the approach of winter and when all other birds are preparing to emigrate from its inhospitable shores south of Palmer Peninsula, or have already done so, the magnificent emperors embark upon that unique course which must always make them the most fascinating of all birds. Deserting the comparative warmth of the sea and the pack-ice, they return to rookeries on the bay-ice and, in at least two localities, on the snow- or ice-covered land. There they lay their eggs, incubate them, hatch out their young and brood them through the Polar Night, enduring in more southerly rookeries 120 consecutive days without sunlight, with temperatures falling below −60 degrees F, and snow-storms raging without cessation for days together.

Moonlight, starlight, twilight or total darkness, but never any sunlight. Days of the moon, when the glaciers are burnished silver and the snow sparkles luminously; when the cold blue sky blazes with brilliant stars, and at noon twilight the northern sky is a crimson flame. Moonless days, when the blue is so deep that it appears black, and the stars are steel points. The snow rings beneath the pressure of the emperors' feet, ice crackles, tide-cracks groan as the water rises, and dull rumblings, rolling cannonades and the roar of avalanches are created by icebergs calving, splitting, capsizing—ten, twenty, thirty miles out in the pack. And, over all, the hanging curtain of the aurora, whose drapery of light reaches to the zenith and falls away in iridescent folds, displaying every possible nuance of luminous yellows and greens. The gossamer veils tremble and undulate, brilliantly lit by powerful searchlights; and beams, with tails of flaming gold, flash up and climb to the zenith in arches of palest green and orange.

Beneath this eternity those few hundred-thousand emperors, in huddles on the bay-ice, on the ice-foot, on the land-ice, fulfil their destiny. One must presume that their ancestors undertook this phenomenal reversal of the normal breeding calendar in an era when the climate of Antarctica was less severe than it is now, as we know it has been, and that the emperors were able to adapt their way of life to a progressively harsher environment. Whatever may have been the climatic conditions when they first reversed the calendar, their kind could not now survive in their present habitat by any other time schedule. The Antarctic summer is too short for the young of such large birds—3½ feet in the skin from beak to feet, and 60 to 90 pounds in weight according to the season—to grow to full size, molt, and be adequately feathered to withstand a first winter in the pack-ice. By contrast, the smaller penguins of Antarctica—the Adélies, gentoos and chinstraps—only three-fifths the size and one-seventh or one-eighth the weight of the emperors, can begin their breeding season in the early summer as late as November, hatch their eggs during the first half of December, and have their young ready for sea late in January or early in February when they are only eight or nine weeks old. Even so, many Adélies must work to a tight schedule because, on returning in October, they may have to walk 50 or 60 miles over the sea-ice to their rookeries to find them still covered with hard-packed winter snow, with temperatures as low as −50 or −60 degrees F. The larger king penguins, only a few inches shorter than the emperors but less than half their weight, also adhere to the normal breeding calendar. Their young, like the emperors', are unable to complete their growth within the requisite period, but they breed only on the islands where environmental conditions are less harsh than on Antarctica, and have evolved a somewhat different method of overcoming this problem.

Virtually all our knowledge of the emperors' extraordinary life history has been gained during the past twenty years, with the opening up of Antarctica by plane and motorized transport. But as long ago as October 12, 1902, Lt. Skelton of Scott's first expedition, when exploring the Ross Ice-shelf, looked down the steep precipices of Cape Crozier from a height of almost 1,000 feet and saw penguins on the ice of a small bay, 5 miles across and 3 miles deep, in a seaward-facing cliff of the ice-shelf—the first breeding rookery of

Gentoo penguins

emperors to be seen by man. Edward Wilson also visited the rookery that spring and twice in the spring of 1903. From that time forward his thoughts were ever upon the possibility of an expedition to Cape Crozier at midwinter, since it was evident, if incredible, that that was the season when the emperors would be incubating their eggs. Even to contemplate the possibility of such a winter expedition was insane, and he realized that it would entail not only traversing about a hundred miles of the ice-shelf's surface in twilight or darkness, with temperatures likely to range from zero down to −50 degrees F, but also the crossing with rope and ice-axe by moonlight of the immense pressure-ridges, 50 or 60 feet high, and the chaos of crevasses between Cape Crozier and the shelf. Moreover, these ridges, which had involved earlier parties in as much as two hours of careful negotiation by daylight, would have to be crossed and recrossed at every visit from his camp to the rookery.

In the winter of 1911, Wilson and his companions, Apsley Cherry-Garrard and Lt. Bowers, actually accomplished this "worst journey in the world," as Cherry-Garrard named it, in the course of a thirty-six days expedition from late June to early August. The surface of the shelf was so bad that they were often obliged to haul one sledge for two or three miles and then return for the second sledge; and on one afternoon, when the temperature had fallen to −61 degrees F, they had to do so by candlelight. By July 6 the temperature was down to the (to them) unheard of low of −77 degrees F, though during a blizzard five days later it rose by 85 degrees. They finally reached the vicinity of Cape Crozier on the morning of July 15, and after taking three days to build an "igloo" of stones roofed with canvas, they went out to the Cape on the 19th. For as far as they could see from the top of the cliffs in the dim light the sea was frozen, and the ice-cliffs were echoing the emperors' harsh metallic trumpeting from the rookery a quarter of a mile or so beyond the pressure-ridges, but they were unable to find a way down through the ridges and, with darkness falling soon after noon, were obliged to return to their crude igloo. The next morning they negotiated the pressure successfully and, guided by the trumpeting of the emperors, located them huddled in a compact group of about a hundred birds at the base of the ice-cliff. They had time only to kill three of the emperors, whose blubber they required for their cooking stove, collect five or six fresh eggs, and note another small group before darkness again drove them back to their igloo. They were not to see the rookery again, nor would anyone else for another forty-six years, for they were storm-bound for the next seven days by a blizzard reaching hurricane force 11 or 12. Today, the rookery can be reached in half an hour by helicopter from the New Zealand and American bases in McMurdo Sound.

To establish a breeding colony on bay-ice several miles from open water, and then incubate eggs and brood young on the ice throughout a polar winter, raises such problems that one cannot conceive how the emperors were able to solve them, nor indeed can we know how they first came to embark upon such a course. Their annual life cycle is in fact wholly related to the formation and dispersal of the sea-ice and, if they have clearly experienced difficulties in adjusting the various aspects of their breeding calendar to the annual and seasonal vagaries of the ice, they have been able to reconcile enough of these to make a success of this unique venture.

How, in the first place, do they determine whether or not a particular stretch of ice is a suitable one on which to establish a rookery? Since, in the weeks before incubation begins, a single rookery may contain upwards of 100,000 closely packed adult emperors, or more than 3,000 tons of penguin, the ice must be stable. It is true that, when subjected to pressure from above, sea-ice does not shatter or crack but gives like elastic, and if the weight is sufficient to press the ice below sea level, the water quickly freezes and raises the level of the ice again. Nevertheless, the emperors show every sign of mistrusting new ice, venturing on to it with the greatest circumspection and, when walking over it groups, usually doing so with one some way in advance and the remainder lagging cannily behind. Mario Marret, the leader of a French party which over-wintered in Terre Adélie during the course of a ten-months study of emperors, has described in *Antarctic Venture* how the first penguins to return to the rookery at Pointe Géologie, 50 miles west of Port-Martin, in the second week of March, assembled initially on a few hundred square feet of old ice which had remained intact throughout the summer, and only gradually moved on to the new ice. This, however, was not very stable so early in the autumn, and the emperors were compelled to change position from time to time, gathering in compact bunches; but despite these precautions, 300 of them were carried out to sea at the end of the month, when a portion of the ice broke away during a storm. Presumably they returned three or four days later when the ice had reformed and become stabilized for the remainder of the winter, as the area of new ice extended further and further over the open sea and attached itself to the old ice. During their absence the number of emperors at the rookery had increased to some 1,500, and the weight of this mass of birds crowded in a comparatively small area caused the ice to bend ominously. Some of the birds thereupon left the main body and went ashore. A little later, when the sea rolled over the sagging ice, the remainder also had to shift; but these spread out from the center of their new position on the ice and formed a large circle, as if aware that the problem could be solved by spreading their 30-tons weight over a greater area of ice.

Once the new ice, on which a rookery has been sited, has become firmly cemented to the land-ice it must, if its foundations are to hold firm throughout the breeding season, be protected in some degree from the disruptive effects of hurricane-force winds and stormy seas, for it must not break up while the emperors are incubating or brooding young chicks, nor before the latter are ready to go to sea. At the same time the rookery ice must not be situated so far from open water that the parents have to walk intolerable distances over the ice in order to obtain food for the young ones when they hatch shortly after midwinter; but it must not become so permanently attached to the land-fast ice that it does not break up in the late summer and enable the young to sail away to sea on disintegrating floes.

The discovery of one emperor rookery after another during the past twenty years or so seemed at first to confirm that the majority are sited on the most stable area of ice on a particular stretch of coast, and that the Cape Crozier rookery was an unsuccessful exception to the rule; for it lies open to the disruptive effects of sea-swell in a locality where every breeze is converted by the configuration of Mt. Erebus and Mt. Terror into storms which periodically blow all the ice out of the Ross Sea and in some years break up the rookery ice at unseasonable dates. The consequence is that the Cape Crozier emperors are sometimes obliged to retreat, while actually incubating or brooding, progressively further away from sea or cliffs, covering as much as a mile in this way during the course of the breeding season. Such a catastrophe had probably occurred shortly before Wilson's winter expedition, when not many more than 100 adults could be counted instead of the expected 2,000. And when his party were eventually able to break camp after the storm and set out on their desperate trek back to base at the end of July, they saw that the hurricane had blown all the ice out of Ross Sea, though what was left of the rookery ice remained intact.

On the other hand, when in 1957 the large Auster

rookery of some 25,000 emperors was first discovered on the sea-ice 35 miles north-east of Mawson on the coast of Mac-Robertson Land, it appeared to be a typically safe rookery, with the additional advantage of being situated only six miles from a large open pool or polynia; for though it was upwards of eleven miles off the land and about eight miles from the nearest group of islands, it was only a quarter of a mile from a line of some hundreds of 100-foot-high stranded bergs. These encircled a large bay and prevented any unseasonable departure of the ice, which was very thick and rough, since it was composed of floes rafted one upon the other when driven against the bergs by pressure from wind and sea, and subsequently cemented together and partially covered with snow. In some conditions the bergs also deflected the full force of hurricanes from the rookery, for when drift-snow, blown by high winds, was streaming across the polar plateau above, obscuring the bases of the coastal ranges and pouring down the sea-cliffs, the air at the rookery itself might be almost calm, and wind strength is perhaps reduced by half at penguin height on the ice. However the relative shelter of its position proved to be deceptive, for shortly after midwinter 1959 the rookery ice was eroded into gullies several feet deep by a succession of severe blizzards, with wind strengths reaching 124 miles per hour. Large numbers of eggs were smashed and deserted, several hundred chicks were killed or buried under the snow, and many adults fell into ice-cracks. Again, at the end of August 1965, there was a fall of ice from one corner of a tilted tabular berg, along the base of which the rookery had been sited that year. The fall killed some of the emperors outright, while others were trapped in ice-cracks and covered with snow which, as it melted with the heat of their bodies and subsequently refroze, encased them in icy coffins. A few weeks later, when a complete section of the berg broke off, its base swung up and smashed the rookery ice. Many of the adults migrated to establish a new rookery around another large berg about three miles distant, but some perished in large cracks, while others, both old and young, were in huddles on numerous broken floes—though the numbers of young appeared to have been greatly depleted.

Shelter from the hurricanes that sweep down off the glaciers to the coasts of Antarctica at frequent intervals must be a major factor in the selection of rookery sites, and many emperor rookeries are, like that at Cape Crozier, situated in the lee of ice-cliffs, despite the ever present danger of ice-falls. It is no doubt because of the high incidence of storm winds that Pointe Géologie is the only locality to have been colonized by emperors on the coast of Terre Adélie, where during the month of March 1951 the wind velocity exceeded 87 miles per hour for 204 hours; while for the autumn and winter months of March, April and May 1912 at Cape Denison further down the coast, Mawson recorded an *average* velocity of 49, 57½, and 60¾ miles per hour respectively, and on May 24 his puffometer registered gusts reaching 200 miles per hour. There is no reason to suppose that his delightful wind-gauge was grossly inaccurate, for a gust of 166 miles per hour has been registered at Hope Bay at the northern tip of Palmer Peninsula. The coastal regions of eastern Antarctica must be the windiest on Earth, and Jean Sapin-Jaloustre, who discovered the Pointe Géologie rookery in October 1950 wrote:

> The wind blows without respite at 100 or 150 kilometers an hour. The blizzard reduces visibility to a metre and lets loose a ceaseless bombardment of small ice fragments; man has great

difficulty in breathing, is incapable of any effort, and is blinded in a minute by a mask of ice. His skin freezes in about ninety seconds. Twenty meters from his hut and he may never find it again.

Nevertheless these winds are often responsible for maintaining wide channels of open water near the coast, and the emperors by taking full advantage of local topographical conditions on the only sheltered stretch of coast at Pointe Géologie (where, incidentally, higher temperatures, reaching 42 degrees F, are experienced than anywhere else in Terre Adélie), have been able to establish what is patently a thriving colony, since it numbers 6,000 pairs or more. Sited on a wide stretch of frozen sea between some small islands and the mainland, it is partially protected from the full force of the blizzards by the hillocks and ridges of the land and by the hummocks and hollows on the ice itself.

The truth of the matter would seem to be that the emperors are not in fact able to select ice that will be perennially stable, sheltered from blizzards, and proof against cliff-falls; but that they play off advantages against disadvantages from year to year, and strike a favorable balance over a period of years. By the late 1950s the Cape Crozier rookery was, for example, much more favorably situated, in a deep enclave behind the main cliff of the ice-shelf, than it had been in Wilson's time. And a few years later was still better protected from blizzards, and completely so from northerly swells, with the result that the ice could remain intact until midsummer. And if the Crozier rookery has the disadvantage of being located more often than not on unstable ice, this is offset by the advantage it gains from being quite close to leads of open water in the ice or to the sea itself. So too, although the similar sized rookery on Beaufort Island, some fifty miles from Cape Crozier, is often beset with pack-ice in the summer, a 3-knot current between it and Ross Island assists in breaking up the land-fast ice in spring and creating a polynia 20 or 30 yards wide between the two islands. In the summer open sea may reach within a few yards of the rookery, which is established on a platform of old ice, 100 or 200 yards wide, at the base of a steep ice-capped cliff.

Incidentally, the difficulties of Antarctic exploration and the inobtrusiveness of emperor rookeries are dramatically illustrated by the fact that the Beaufort rookery was not discovered until 1966, although its position in the south-east corner of the Ross Sea has been right in the path of expeditions for more than sixty years. Likewise, it was not until 1959 that the immense rookery of some 50,000 pairs of emperors on Coulman Island was discovered, though this island is also in the Ross Sea and only about 150 miles from the long known and very large Adélie rookery at Cape Adare.

A further example of an unfavorable situation being compensated for by easy access to fishing waters is provided by the emperor rookery on the Dion Islets at the mouth of Marguerite Bay on the west coast of Palmer Peninsula. This small rookery of but 150 pairs is one of the only two known to be sited on terra firma, and covers approximately 700 square yards of low-lying shingle beach on which snow is firmly packed down to produce a layer of ice about one foot thick over the stones. It suffers the disadvantages of being exposed with very little shelter to the prevailing strong northerly winds, and of probably being awash in the summer after the protective barrier of sea-ice has broken up. But, despite the fact that the islands are surrounded by pack-ice

4 or 5 inches thick for seven months out of the twelve, its occupants enjoy the advantage of being able to fish in open water only a few hundred yards from the rookery early in the spring, because reefs and strong currents engineer the formation of polynias before the fast-ice breaks up.

Adélie rookeries are also sited to take advantage of polynias, as in the case of two on the coast of Terre Adélie, where polynias are formed in the sea-ice during the early summer by warm melt-water draining from capes and islands. The occupants of the rookeries are thus able to obtain krill close at hand instead of having to traverse 50 or 60 miles of ice to the open sea. So too, the 500 pairs of Adélies at Cape Royds —their smallest known rookery—are able to breed that far south because, although the sea is frozen when they return from wintering in the pack-ice, strong offshore currents assist in keeping the east side of McMurdo Sound free of ice. By the middle of October a large polynia in which they can fish has usually formed off the cape, while by the time the chicks hatch, krill can be obtained in the open sea less than a mile from the rookery.

On the other hand a rookery enjoying the advantage of stable ice may suffer the balancing disadvantage of being situated at a distance of 60 miles or more from the seaward edge of the ice. Emperors from some perhaps at present undiscovered rookery on the west coast of Palmer Peninsula have been encountered walking over the ice among the offlying islands at least 70 miles from the sea. Many Adélie rookeries are also situated 30 or 40 miles from the sea, and one has indeed been reported to be 200 miles distant. This would appear to be highly improbable, for if in exceptional years the sea off Cape Royds remains frozen until the end of December, the breeding cycle of the Adélies is disrupted:

they are obliged to trek 60 miles over the ice to open fishing waters, with the result that mates are unable to return in time from fishing expeditions to relieve partners brooding chicks, and the latter are abandoned to starve. An Adélie, with its very short legs, has a stride of only 4 inches and walks at a rate of 120 steps a minute, though proceeding more swiftly when tobogganing. A distance of 60 miles would therefore entail a trek of at least five days, if made direct as the skua flies, and providing that it did not include numerous detours around hummocks and ice-cracks, and numerous and prolonged halts to rest and "gossip" with penguins going in the opposite direction. Possibly catastrophes have occurred fairly frequently at Cape Royds, for an extensive residue of guano over a wide area at the cape suggests that the rookery may formerly have been much more widespread, and comparable perhaps to those on the other side of Ross Island, where the colonies at Cape Crozier and Cape Bird total hundreds of thousands of birds.

How do Adélie penguins and others, returning from wintering in the pack-ice, orientate through the chaos of featureless pressure-ridges to rookeries situated far from the sea? A minority perhaps are not able to find them, though in this respect it must be said that penguins in general are obsessed with the pleasures of "walk-abouts" and when not actively engaged with eggs or chicks are extremely sociable. If one unemployed king penguin sets off on a walk-about he is joined by a number of other unemployed, and both emperors and Adélies wander around in bands. Adélies, in particular, tramp down broad beaten roads in the snow around inland lakes, and less trodden tracks to the summits of hills. Small groups of chinstraps are in the habit of ascending almost to the lip of an old crater on the Orkney island of Southern Thule. There is indeed a report of chinstrap tracks 250 miles from the sea. The tracks of an emperor have been sighted at the end of December, 240 miles inland from the seaward edge of the Filchner Ice Shelf and 300 miles from the nearest known rookery, and those of an Adélie penguin 186 miles in from the coast of Marie Byrd Land. Both sets of tracks led away from the coast.

Only a few experiments, designed to test a penguin's homing powers, have been carried out. Adélies, removed to distances of up to 700 miles from the Cape Crozier rookery and liberated, after plane flights of several hours, on the featureless wastes of the interior, set off when released, not on a direct bearing for the rookery but NNE towards the coast. No matter what time of day they were released they invariably adopted this bearing, despite the constantly changing angle of the sun to it, though when released under cloudy conditions they either settled down to sleep or dispersed in varying directions, reorientating however on to the coastal bearing when the sun reappeared. Apparently all returned to the rookery in good condition. Moreover five male Adélies, which had been unsuccessful in breeding at Mirny rookery and were flown 2,375 miles to Cape Crozier, set off, when released, not on the NNE coastal bearing, but on a NW bearing which would have taken them to the sea in the Mirny area. Nevertheless, two and probably three, of these Mirny Adélies were recovered in their rookery ten months later during the following breeding season. From these limited experiments it would appear that a penguin is able to make a correct assessment of its bearings wherever it may be, and that by orientating towards the coast, rather than directly to the rookery, it will be able to obtain food in open waters before reorientating to its rookery or to winter fishing grounds. The primary navigating instrument would appear to be the sun. And if this is the case a penguin must be equipped with a form of "biological clock" which assesses the changing position of the sun during the course of the day.

Although only the emperor penguins establish their rookeries on sea-ice, with all the problems this entails, other species, nesting on terra firma, also have difficulty in selecting suitable terrain. The whole structure of the Adélies' nesting cycle rests, for example, literally on the presentation of pebbles during the days of pair forming and courtship, and these will subsequently serve as foundations for their nests, raising them above the channels of melt-water. Similarly, the gentoo penguins prefer to nest on the tussac, whose mounds rise above the puddles and the elephant seals' wallows, and are also clear of snow earlier in the spring than is the ground below. This preference obliges them to change the sites of their rookeries every two or three years, to allow thc tussac to recover from their incessant trampling.

An Adélie rookery must therefore be situated where quantities of pebbles and larger stones are available. But the evidence is conflicting as to how greatly they are further influenced in their choice of sites by the latter's freedom from drifting snow. Some observers have reported that a colony may lose half of its eggs in a blizzard, and that the birds themselves are liable to be smothered in drifts accumulating in sheltered hollows. Herbert Ponting, on the other hand, noted that though all the incubating Adélies in the Cape Royds rookery disappeared under knee-deep snow early in December, none deserted their eggs, despite the fact that thin films of ice covered some of the air-holes opened by the birds' body heat. About an hour after the storm had ended, the blanket of snow began to flatten down in the sunshine, and the tips of the Adélies' beaks appeared. Although the sun's heat, together with their own, melted the snow into slush which soaked the eggs, only a few of the Adélies deserted at this stage. And though this storm was followed two days later by another blizzard, which put down a waist-deep pall of snow over the rookery, chicks subsequently hatched from hundreds of the nests. But whatever effect the incidence of snow may have on the selection of rookery sites, the majority are certainly located on rocky slopes exposed to the winds. When Edward Wilson landed at Cape Adare in January 1902 he found "millions" of Adélies covering not only some 200 acres of plain and hillside, but extending in small colonies over the summit of the headland at a height of more than 1,000 feet. Today, it is estimated that about half a million Adélies cover 500 acres of Cape Adare. Their rookeries spread not only over the flat water-worn rocks near sea-level but over the steep faces of cliffs that are almost inaccessible to climbers. There are forty or fifty pairs, together with a few chinstraps, on the one part of Peter I Island's 70 square miles which is not heavily glaciated, despite the fact that this oceanic island is usually surrounded by dense pack-ice. And there are a dozen or score of pairs, among the macaronis and chinstraps, on the small and almost entirely ice-covered island of Bouvetøya, where the limited area of flat ground available for a rookery is constantly threatened by avalanches and rock-falls, and where the penguins have to compete for living space with fur seals and elephant seals. The size of an Adélie rookery appears to be restricted only by the extent of suitable ground available.

Although the South Shetlands and Orkneys suffer super-severe gales comparable to those striking eastern Antarctica, the Adélie rookeries on those islands too are situated on the windiest and most exposed slopes, which are swept free of snow and are also surfaced with quantities of the essential small stones. Despite these climatic disadvantages, to which can be added that of heavy snow-falls from persistently overcast skies and engulfment in the pack-ice for eight or ten months of the year, immense numbers of Adélies and chinstraps breed on the Orkneys, including some 5 million Adélies on Laurie Island alone. As an indication of the superabundant wealth of food in Antarctic seas, one zoologist has made a rough estimate that the Adélies on Laurie Island catch about 9,000 tons of krill and small fish every day within ten miles of the island—the equivalent of the catch of a fleet of seventy modern trawlers with landing facilities ten times greater than those of Aberdeen.

7: Emperors in the Polar Night

On a day in the second week of March, shortly after the new sea-ice has formed, the first emperors return to the rookery at Pointe Géologie. No doubt they return two or three weeks later to rookeries further south. Mario Marret has described how shortly after the first two emperors had arrived, a third came in from sea, clambered on to the ice, stood upright and shuffled towards them. On reaching them he bowed, lowering his head in the formal emperor display, eliciting a similar response. The three then stood face to face for a while, exhibiting their sumptuous plumage. The midnight-blue of their backs and their white satin bellies glittered with crystals of ice and salt; jet-black skullcaps, reaching down to the nape and extending over cheeks and chin, were set off by the golden or copper patches on the sides of their necks, and by the rose-colored trimming, deepening to violet, below and on either side of their lower mandibles. At first the emperors arrived in small groups, with the mass returning in parties of up to 300 during the last ten days of March. By the end of April when, with sixteen hours of darkness and the temperature 30 degrees below zero, the sea had already frozen to a distance of more than sixty miles off the land, the rookery had its full complement of 12,000 or 13,000 emperors spread over some five acres of ice. Marret goes on to describe how pairs were formed and reformed by means of a vocal display, since the sexes are (to the human eye) similar in appearance, posture and behavior. On arriving at the rookery a file of emperors would disperse and, coming to a halt, would crane their necks and raise their beaks as if listening; then they would rub their heads against the upper part of their flippers, first to one side and then to the other. Finally one would lower his head, fill his lungs and begin to "sing." Opening with a series of runs on a low register, these would rise in semi-tones, grow shriller, and be resolved on one sustained vibrant note. Having concluded his song, the bird raised his head and stood motionless for a while in the listening attitude. If he received no reply, he moved further into the rookery and repeated the performance before another group—as some might have to do again and again for several days and yet in the end be unsuccessful in finding mates. But in the normal course of events a male eventually received a reply from a female, and after facing each other in silence for a short while, one or other lowered its head and began to sing, eliciting an immediate cooing response. However, even at this stage, a pairing might not have been finalized, for after the performance had been repeated several times one or other might break away and wander off to another part of the rookery. The display of the colorful neck-patches by bowing may set the seal on a mating, for when these have been painted out experimentally, the affected emperors have been unable to secure mates. But many no doubt are old partners, as has been demonstrated by ringing among smaller species of penguins.

Once pairs have been formed, and the time approaches for the females to lay their single eggs in the first week of May and on into June, the emperors are confronted by another problem. It is now approaching midwinter and the extent of the 5-foot-thick ice between their rookeries and open

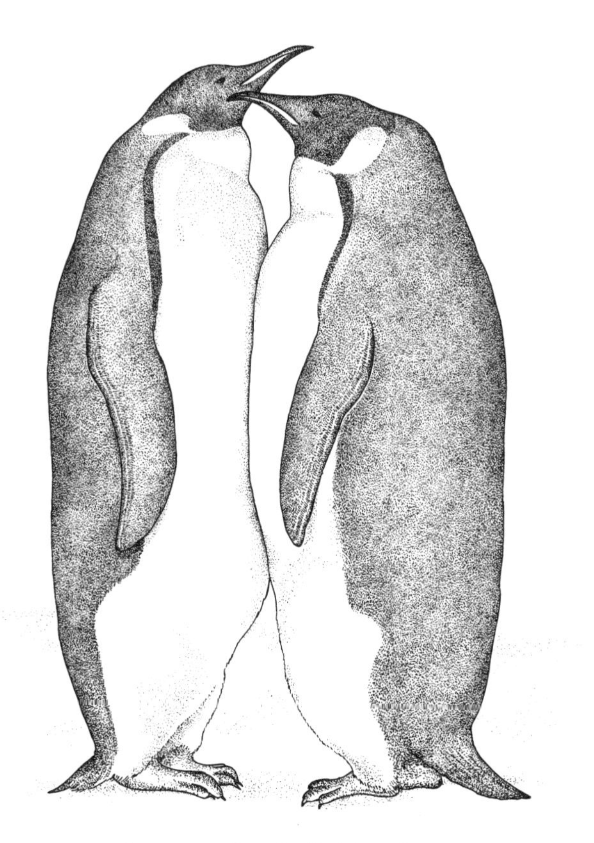

Warbling to her mate, female (left) and male emperor penguins

water is becoming ever greater. It is also the season of blizzards and of almost total darkness during cloudy moonless spells. Clearly, under these conditions, short alternating fishing trips by whichever partner is not incubating are not feasible, particularly at those rookeries scores of miles from the nearest open water, for these would entail long walks in sub-zero temperatures, resulting in loss of fat difficult to replace under winter conditions. Although emperors can travel faster than a man can run by tobogganing on their breasts, propeling themselves with alternating strokes of flippers and powerful feet, this technique is only practicable for relatively short distances over smooth ice, and also in negotiating narrow leads which they are unable to jump across. The only solution to the problem is for one emperor to undertake the entire incubation while its mate migrates to sea, returning with krill or fish at the time the chick is due to hatch. Since the female, who is not so heavily blubbered as the male, has by this time already lost nearly a quarter of her weight as the result of two months starvation at the rookery, and has also had to produce an egg 5 inches long and between ¾ and 1¼ pounds in weight, she it is who entrusts the eggs to the male and leaves him to perform this "heroic" and unique incubation through the long polar night. Since in most species of penguins the females are reported to delegate the initial spell of incubation to the males, despite the fact that all except the emperor enjoy the advantage of long hours of daylight, this procedure was presumably evolved as a means of overcoming this problem of the distance between rookery and fishing waters, for among other penguins the duration of the male's spell of incubation appears to depend on this factor. Male kings, for example, may incubate for the first 2 or 3 weeks of the 7 or 8 week period; male Adélies for as little as 2 days or as many as 18—half the total incubatory period—though, even so, some may have to fast for a month, if one takes into account the time involved in trekking over the ice on their initial return to their rookeries, and their subsequent courtship and nest building activities. On Heard Island, however, it is apparently the female macaroni penguins who undertake the 5 weeks' incubation before being relieved by the males, who return in time for the chicks' hatching and continue to brood them for 3 weeks while the females supply the food.

Curiously enough, although it has always been taken for granted that it is in fact the male emperors who normally undertake the incubation, it would seem that this has been confirmed by examination in only about a dozen cases. Some females, deserted by their mates before being able to transfer their eggs to them, may remain to incubate. A proportion of those females who go to sea must die during the winter from one cause or another, just as some eggs are lost by incubating males. Females returning to mates without chicks may therefore take over those of males whose own mates fail to return.

After she has laid her egg, the female emperor parades to and fro, stopping from time to time to disclose the egg to the male following her. Finally, holding her flippers outstretched and distending her belly, while craning her head and swaying from side to side, she begins to mark time, trampling heavily; then, opening her feet, she allows the egg to roll on to the ice. Immediately the male hooks his beak over it, rolls it between his own legs, and clumsily and with great difficulty, succeeds in hoisting it up by drawing his feet together, and tucks it away between the upper part of his legs and the lower part of his heavy and protruding belly. This accomplished, he begins to sing, and the female soon sets out on short walks through the rookery. The male endeavors to follow her, but, encumbered by the egg, is unable to keep up, though a few males apparently start out with the intention of walking to the sea with their eggs, or else abandon them. Some time within the next twelve or forty-eight hours the female emigrates, alone or with two or three other females, to the sea, which may now be as much as 125 miles distant.

And so the male is left to his nine-week vigil. No part of the egg, protected by a pendulous fold of abdominal skin, touches the ice, while the bare vascular brood-patches and surrounding area of feathered skin transfer heat from his body to the egg. He is indeed a most efficient incubator, for when the air-temperature is as low as −15 degrees F the internal temperature of the egg is +88 degrees F. When he moves about he does so with the greatest care, shoulders hunched, body taut, nails hooked firmly into the snow, and he walks wherever possible on established trails trodden down by non-incubating birds. Even if he falls over or sinks down on his belly to toboggan he may not necessarily lose or smash the egg, and he can indeed retain a firm hold of it while scratching the back of his head with the claws of one foot. The large size of their eggs does not prevent the males from moving quite long distances. Under favorable weather conditions during the earlier stages of incubation, if not subsequently, small groups may be seen shuffling around on the ice as much as a mile from the rookery, with each emperor still carrying his precious egg; and if threatened by ice-cliff falls or by a break-up of the rookery ice they can, as we have already noted, remove themselves and their eggs to a new site. G. M. Budd, a member of the Australian National Antarctic Expedition, has described in *Animals* how one afternoon, when the glow in the northern sky had almost waned and the first pale rays of the aurora were weaving and swaying over the polar plateau, he and his companions finally located the Auster rookery after a six-day search through a vast chaotic waste of ridges and dunes of snow and tilted blades of ice. At first the lunar silence was broken only by the slow creaking of the snow beneath their tread, and by an emperor's musical trumpeting echoing around the ring of icebergs. But when they approached near enough to the rookery for the emperors' overall blackness to be relieved by the light-colored patches around their ears, which showed as golden smudges against the blue ice, they became aware of a strange sound, as if a thousand teacups of the finest bone china were being gently knocked together. "We stood spellbound," Budd wrote, "while this faint, but quite clear sound continued and the bowed black figures shuffled slowly away from us. Later we decided that it must have come from the eggs moving on the penguins' horny feet as they walked—or possibly from the hard-packed snow under them."

How did the emperors come to incubate in this peculiar manner, rocking on a tripod of heels and stiff tail-feathers? Small penguins, which lay two eggs, also tuck these between the upper side of the tarsii and their brood-patches, but incubate in a crouching posture. If however, as has been conjectured, the emperors first took up residence on Antarctica when the climate was less severe, the habits of king penguins are not without relevance. Unlike the emperors, the kings' preference for a rookery site is the largest available area of snow-free ground, such as a bare stony moraine a few hundred yards in from the sea, though in most instances there will be a glacier or permanent snow-bank hard by the rookery, from which the birds can drink and on which they can bask on calm sunny days. At Paul Beach in South Georgia's Bay of Isles a raised beach slopes gently up to the base of a cliff 200 yards from the sea; two headlands of the cliff enclose an embayment half way along the beach. Here the rookery is sheltered by cliffs on three sides, and is sited on a higher sloping raised-beach providing a cobbled and comparatively well-drained floor. The landward half of the beach is, however, covered with dense tussac, containing the wallows of elephant seals, and when heavy rains churn the tussac into morasses and lakes, the penguins' winding paths become almost impassable. Moreover, throughout the summer, snow-melt from the rolling tussac above the cliffs flows through the rookery in two shallow streams which drain into a large muddy pool. During the winter the streams dry out, the pool freezes, and extensive snow-drifts, forming below the cliffs, gradually encroach upon the floor of the rookery, which is overlaid by 3 feet or more of well trodden impacted snow. From this description it will be seen that the king penguins incubate under very different conditions to the emperors. Nevertheless, their technique is similar. On Macquerie Island, for example, Edward Wilson found thousands of kings incubating on flats of loose stones. Water, trickling over the flats from a stream flowing down a valley, contributed to churning the rookery into a vast pig-sty of liquid mud and guano; but the kings protected their eggs by supporting them on their insteps, with toes raised off the ground, and covering them with the loose hanging lappets of skin and feathers, just as the emperors do on the snow and ice.

Having solved the problem of preventing their eggs and the embryos of the next generation from becoming chilled, the male emperors had also to solve the problem of how they themselves were to survive the polar winter, huddled in the company of hundreds or thousands of their kind in the lee of ice-cliffs or far out on the bay ice. Before even beginning to incubate they have already fasted for eight or nine weeks, surviving on their reserves of subcutaneous fat based on oils extracted from their rich planktonic and fish food, for the large quantities of snow they consume do no more than slake their thirst. An inch or an inch and a half thick, this overall layer of blubber amounts to as much as one-half of their body weight of upwards of ninety pounds when they first return to the rookery. For body heat they, like other penguins, must rely partly on their blubber and partly on their plumage which, being composed of a "chainmail" of seventy oily feathers to every square inch, is resistant to ruffling by the wind. Moreover the feathers, being quite stiff and turned over at the tips, overlay and trap a layer of air—a bad conductor of heat—next to the skin, while after-shafts or fluffy outgrowths near the base of the feathers provide additional insulation. So efficient are these various insulators that, when the temperature rises to near freezing point, Adélie penguins evince signs of heat distress, ruffling their feathers and, in so doing, allowing the insulating layer of air next to the skin to break up and some of the body heat to escape, or holding their flippers away from their sides and exposing additional areas of body surface to the air.

Nevertheless, this efficient insulation does not in itself provide the incubating male emperors with sufficient protection against the uniquely low temperatures and penetrating blizzards of the Antarctic winter. Their reactions to the "discomforts" of a polar winter are perhaps not very different from those of a man, who has little difficulty in keeping actively warm when the temperature is as low as −40 or −50 degrees F, provided that there is no wind. Many years ago, Frederick Schwatka, for example, recounted in *Nimrod in the North* how, when trotting forty miles after a herd of caribou in the Canadian Arctic, he found himself too warm on a windless January day, despite 100 degrees of frost. But, when unable to generate heat by activity, even a light breeze with a temperature of zero F produces conditions of extreme discomfort for both men and incubating penguins

—though penguins probably do not suffer from cold feet, since these contain a minimum of blood-cells. Extreme cold the emperors counter by a remarkable communal cohesion. If the temperature falls to around 15 degrees F when a strong wind is blowing, or to below zero with light winds, the emperors fall back on their last line of resistance and huddle together in small groups. As early in the winter as April, twenty or thirty huddle up on cold nights of 60 degrees of frost; and by the end of May an entire colony of incubating males, whether it numbers 1,000, 5,000 or 12,000, will pack together as closely as possible on forecasting a blizzard. Facing, with heads bowed, towards the center and leaning slightly inwards with beaks resting on the shoulders of those in front, their arched backs present a compact oval-shaped mass—comparable to the testudo formed by the shields of Roman legionaries—against a blizzard. And so they remain in almost total silence for days at a time if the storm is prolonged, or during a period of low temperatures associated with strong winds. But within the ranks of the testudo there is a constant displacement of individual penguins, as birds in the outer ring on the windward side begin to shiver with cold and, forcing their way through those in front, shuffle in towards the center, to change places with those in inner rings. Thus—if their behavior has been reported correctly—every member of the testudo gains the opportunity to stoke up heat and energy periodically within the warming protective shield of his fellows. Isolate an emperor from the testudo when the air-temperature is −24 degrees F, and he very soon begins to shiver violently. And experiments indicate that any penguin which failed to huddle would not only be liable to lose more than 80 per cent of his body heat, but would also lose weight twice as fast as his fellows, and would probably be unable to complete the full incubatory period, almost half of which is passed in the testudo. Incidentally, the formation of the latter is possible only because, unlike the smaller penguins, emperors are peaceable birds. The activities of those pushing in from outer rings provoke only mildly retaliatory lunges with beaks. Nevertheless, eggs dropped during the formation of testudos, to be frozen or crushed, may account for up to 70 per cent of a rookery's total egg losses. After watching the large Pointe Géologie colony of emperors huddled in a testudo for protection against the stinging particles of ice and snow driven by a 75-mile per hour blizzard, one of Mario Marret's party, Jean Rivolier, drew a macabre comparison between their behavior and that of the ill-clad undernourished inmates of Mauthhausen concentration camp during the bitter cold of a Central European winter. Forced to remain outside their huts for hours at a time the men would begin by standing together in closely packed groups, and subsequently circle slowly round and round, so that every man obtained in turn some protection from the full force of the wind, together with the opportunity of rewarming himself, or at least of not freezing to death—"adopting a technique of the glacial era in order to survive the civilized twentieth century," was Marret's comment on this comparison.

As the weeks pass and their reserves of fat decrease, the emperors' tendency to huddle increases with the slowing down of their metabolism and resulting lethargy; and the British ornithologist Bernard Stonehouse observed that although at midday on July 18 a small group at the Dion Islets rookery turned towards the bright light of the returning sun when it appeared for a few moments above the horizon after its two months' absence, the main mass slept on undisturbed in their testudo.

8: Crèches and Predators

It is early in July, during the season of the most terrible blizzards, and when there is still less than an hour's twilight at midday, that the first of the emperor chicks hatch at Pointe Géologie, three or four weeks sooner no doubt than in rookeries further south. Their shrill whistles pipe from all quarters, and their hatching is responsible for a spectacular rise in activity in the previously somnolent rookery. Testudos may now be formed only in extremely severe weather, and not always then; for even in a 90-mile per hour gale, with the temperature as low as −25 degrees F, the adults may do no more to protect themselves than group closely together and turn their backs on the storm, closing their eyes or covering them with their nictating membranes against the stinging drift.

Ideally, the female emperor is, by some extraordinary physiological adjustment, able to time her return to the rookery after an absence of between seven and nine weeks, to coincide approximately with the hatching of the chick. At Pointe Géologie the earliest returning females have indeed regularly anticipated this event by periods varying from a few hours to fourteen days. Their return has as electrifying an effect on the males as does the first appearance of the chicks. The earliest female to return to the Dion Islets rookery did so on July 19, twelve days before the majority of her fellows, although the first egg hatched on this day; and Bernard Stonehouse describes in his paper on the emperor penguin (FIDS scientific reports) how when she tobogganed in at high speed over the wind-packed snow apparently highly excited, and uttered a short interrogative cry, a four-day-old huddle of males broke up immediately. Stonehouse believed that a female returns to any male who will surrender his chick or egg to her, and not necessarily to her original mate. He observed that a newly arrived female would push steadily against the chest of a male, who would either refuse to release the chick, turning away and maintaining his hold of it, or would step backward, leaving the chick lying on the snow. In the latter event the female would scoop the chick up and hold it firmly on her feet while the male would stand beside her, apparently reluctant to desert his charge. Male kings display a similar reluctance, staying with their mates for two or three hours after the changeover, and not finally leaving the rookery for sea until the following morning or forty-eight hours later.

Mario Marret, on the other hand, asserted that a returning female might have to "sing" for hours, while making her way among the thousands of males, before her voice was ultimately recognized by her mate. And he describes how it was at this time, in those rare intervals when the everlasting storm-winds mysteriously quietened for a brief space according to no laws known to man, that one would hear from afar the emperors' orchestra. "A symphony of squawking, quacking and chattering," he wrote, "from the deeper trombone of the male to the muted trumpeting of the female, through alto, piccolo and flute with occasional silences for a solo followed by a chorus as though of lamentation: a symphony for wind and strings in the atonal scale."

There is clearly a conflict of observation or, possibly, a variation in individual behavior here.

Marret's description of the returning females might seem to be supported by Stonehouse himself in his account of the behavior of female king penguins at Paul Beach, South Georgia:

> On her return from sea the female usually spends one or two days on the beach before relieving the male. She may sleep in a group of molting birds, wander aimlessly with other birds in the tussock grass, perhaps respond several times to advances from courting birds before finally making her way to the edge of the rookery. Standing at the edge of the breeding group in which her partner is incubating, she gives several long calls in succession. In spite of the general noise of the rookery, the partner usually hears and recognizes the call immediately; he responds by stretching upright and giving his own calls, which in turn are recognized by the female. The female then beats her way through the group of incubating and courting birds, apparently in the direction from which her partner was heard. She may stop two or three meters short of the breeding site, or may overshoot, passing the male and perhaps attacking him or being attacked by him on the way. She calls again, is re-directed by his response, and finally stands beside him.

At birth the 6-inch-long chicks weigh only about three-quarters of a pound and are almost naked, perhaps because a covering of down would impede the transference of heat from the parent's brood-patches. At a later stage, however, they grow a dappled silver-gray down, which lengthens and thickens into a weather resistant, almost fur-like plumage. The extraordinary fact is that, if a female's return is delayed, the male can provide food for the chick temporarily by secreting a whitish mucus somewhat resembling "pigeon's-milk" and containing proteins, fats and sugar, from a thickening in the wall of his gullet, despite his long fast which has reduced his weight by half. The female, for her part, begins to regurgitate triturated krill or fish to the chick immediately she has gained possession of it. She may indeed attempt to do so when it is only half out of its shell or even when it is piping within the egg. The cropful of seven pounds with which she returns is sufficient to furnish the chick with small meals until the male has had time to make the long trek over the ice, now at its most extensive, to fishing waters, feed himself up again, and return with further supplies after an absence of three or four weeks. At some stations, or perhaps at some seasons, the emperors feed mainly on squid. These they would be able to capture in surface waters at night, though experiments have demonstrated that an emperor can in fact dive to depths of 180 feet and remain submerged for as long as 18 minutes.

For six or seven weeks after birth the chick, standing on the feet of whichever parent is not away fishing, is held tenaciously between brood-patches and tarsii. This is one of the critical periods of its adolescence. So strong is the reproductive urge of the emperors—what else could impel them to breed at the wrong season of the year?—that any parent that has lost its chick will brood a dead one, just as one that has lost its egg will incubate a chunk of ice. Every live chick straying an inch or two from its parent's "pouch" risks being torn to pieces by being seized at either extremity by bereaved parents, desperate to nurse it themselves; and upwards of 6 per cent of the hatch are in fact lost in this way.

By the end of this period of brooding, late in August or early in September, the chick is too large to squeeze into its parent's "tent," but is now sufficiently protected by blubber and down to maintain its own body heat. This

Emperor penguin with month-old baby on feet

development enables both parents to go off fishing together, an essential change because, while the rapidly growing chick now requires large quantities of food, inshore waters are still covered with ice 5 feet thick, entailing long treks to the open sea for those emperors from "inland" rookeries with no open pools or polynias in the more immediate vicinity. As more and more parents temporarily abandon their chicks on these fishing expeditions, so the latter are liable when disturbed to form small loosely knit groups in the center of the rookery, with a ring of any not otherwise occupied adults around them. Subsequently the chicks coalesce in crèches of 1,000 or more. Crèches are also formed by young Adélies and by young king penguins. And it was formerly believed that the chicks in a crèche were fed indiscriminately by any parents arriving with krill or fish or squid, and that only the fittest or most aggressive of them would survive, because in the pell-mell rush towards any returning adult many of the smaller and weaker ones would never receive any food. All the pioneer naturalists in the Antarctic stressed the turmoil in a rookery when an adult returned with food, and Edward Wilson observed that adult Adélies returning to a rookery (as opposed perhaps to a crèche) were so pestered by the chain of chicks through which they had to pass in order to reach their own that, as often as not, they ultimately disgorged to a strange chick and not to their own. However, the more prolonged observations of better equipped contemporary naturalists indicate that in many instances the returning parent, whether emperor, king or Adélie, stops on the fringe of the crèche and calls out its own chick to be fed. Although the crèche may contain 2,000 or 3,000 tightly packed, sleeping chicks one head shoots up with a piercing whistle immediately the parent calls, and its owner begins to push its way through the ranks of its uninterested companions. Calling repeatedly, parent and chick "home" on their respective whistles, though they may walk past each other and still be calling when only a few inches apart before ultimate recognition, for penguins are apparently near-sighted. If the parents of a particular chick fail to return, it starves, despite the fact that it is surrounded by adults disgorging food to their own young.

It is difficult to reconcile these totally opposing observations, but undoubtedly considerable confusion must occur on the arrival of adults with food at rookeries containing hundreds or thousands of chicks. What must often happen is suggested by the experience of Francis D. Ommanney who noted, in *South Latitude*, that, though parents returning with krill to groups of a dozen or more Adélie chicks on Deception Island would call to the latter—whereupon these would leave their group and run to the water's edge—the parents would sometimes disgorge to the wrong chick. He and Jean Prévost, the ornithologist in Mario Marret's party, also observed that in the latter part of the summer the parents of both Adélies and emperors would become bored with the feeding routine, the family bonds would loosen, and parents would disgorge to any chicks that importuned them.

A crèche usually includes a few adults, and it was also formerly believed that these acted as guardians, protecting the chicks in the crèches from such predators as skuas and giant petrels. But it is now believed that the adults' presence is fortuitous, and that they do not play the role of protectors. However, since they intermittenly chase away marauding skuas and petrels, they are certainly beneficial, and the evidence is, again, conflicting. Marret, for example, observed that young Adélies in a crèche unattended by "guardians" invariably panicked when a skua flew overhead, but did not do so when adults were present. And other naturalists claim that the crèche serves as an anti-predator deterrent, whether or not adults are present, because skuas never attack any chick which is in one. There is indeed confusion as to the degree of predation to which penguins are subjected. On the one hand there are those who assert that a young penguin is not capable of warding off the attacks of a skua or a giant petrel or Dominican gull, all of which are for ever patroling the rookeries. This is confirmed by the observations of several naturalists who have watched skuas dragging young Adélies almost as large as themselves from the rookeries, or squatting on their backs when their parents were away fishing, and pecking out their eyes preparatory to killing them. In the South Orkneys the brown skuas swoop incessantly over the rookeries of small penguins, snapping up eggs and chicks from unwary parents. And on South Georgia they feed almost entirely during the breeding season on the eggs and chicks of the gentoos, and on small abandoned king chicks—though groups of larger young kings will chase them about the rookery.

On the other hand there are those who assert (improbably) that a skua is incapable of killing an Adélie chick more than three weeks old, ie when it is barely large enough to join a crèche. More credible is the possibility that since on the more southerly coasts of Antarctica the McCormick's skuas begin to breed much later than the Adélies, the rookeries of the latter cannot provide much prey after the middle of January, at that season when the Adélies are abandoning their rookeries just when the skuas' own chicks are becoming increasingly rapacious. A study of McCormick's skuas near an Adélie rookery on Ross Island indicated that they could secure almost enough food for their chicks from other sources. In any case, since each pair of skuas establishes a hunting territory over a rookery, those unable to do so will have to obtain their food elsewhere. Both Antarctic and snow petrels are, for example, harried and forced to disgorge on the wing, though one would not have supposed that their crops contained much of value to a skua. It may also be the case that young emperors are not much troubled by skuas, since the latter do not return to Antarctica from their winter wanderings over ice-pack and ocean until the middle or end of October, by which date most young emperors are certainly too large to be killed by them, though Jean Rivolier states in his book *Emperor Penguins* that some skuas return at midwinter to Terre Adélie and scavenge on the dead chicks in the emperor rookeries.

The subject of predation at penguin rookeries is clearly one that requires more study, as does the effect on the other Antarctic avifauna of these huge populations of penguins. Giant petrels, those collossal fulmars with an 8-foot wing span, are, for example, reported to be the chief predators at the rookeries of both emperors and kings, though they do not breed near the more southerly emperor rookeries. Jean Prévost found that they were responsible for a mortality varying between 5 and 34 per cent at the Pointe Géologie rookery, ripping up the young emperors with their massive beaks, which are powerful enough to crunch up the bones of young penguins; while on South Georgia groups of a dozen giant petrels visit the king rookeries at first light almost every morning throughout the winter from late May to late August, and are again responsible for considerable mortality among the weaklings. But this predation by giant petrels must be considered beneficial rather than harmful, for since they can only shuffle along for a few feet at a time, maintaining their balance with half-open wings, before sinking on to their breasts, their victims must necessarily be the weaklings. If a giant petrel makes a short rush at king chicks herded tightly together in a crèche, the latter have no difficulty in evading it. But if their ranks open to disclose a weakling or fallen chick, it is immediately snapped up. Giant petrels may constitute a more serious menace to penguins when the young ones are entering the sea for the first time.

Another predator at the rookeries, though probably rated as only a minor nuisance by the penguins, is the white sheathbill or kelp bird. Known to the Norwegian whalers as the *rype* or ptarmigan, it has been described as having the appearance, gait and flight of a pigeon and the beak and voice of a crow. With the exception of the pipit, the sheathbill is the only Antarctic bird whose feet are not webbed; but despite the additional handicap of apparently weak powers of flight, it is often encountered during the non-breeding season among the pack-ice, resting on floating kelp

Great skua grabs an Adélie penguin egg as penguin hen screams in protest

or logs, several hundred miles from the nearest land, though never north of the Polar Front. It is a scavenger on the excrement, placenta, umbilical cords and carcasses of seals, and especially on pickings in the penguin rookeries, including the occasional egg and possibly very young nestlings. A score or two of sheathbills may apportion their time during the summer between a rookery of kings and the algae along the tide line. Whether sheathbills could exist on the Antarctic islands and the north of Palmer Peninsula if there were not in the first place krill, and in the second place penguins feeding krill to their chicks, is a nice point. Apparently the sheathbill's own chicks cannot be reared without an assured supply of krill, and this the parents cannot obtain by their own efforts. Their breeding calendar has therefore been arranged so that when their chicks are of a size to require the maximum amount of krill, the young penguins are also receiving large quantities, of which the parent sheathbills pick up the spillings. Some indeed deliberately harry such large penguins as kings, when these are regurgitating food to their chicks, by flying up at their heads as the parent is in the act of passing food, and flapping their wings hard against the two penguins' interlocked beaks. The invariable result is that both withdraw their heads in order to lunge at the marauder, and the food splashes on to the ground, to be picked up and carried away by the sheathbill to its own young. A few skuas also obtain food from penguins in this way.

But, to return to the subject of crèches, the most advanced form has been developed by the king penguins whose young, hatched at midsummer are, as we have seen, unable to complete their period of adolescence within the time available, and must therefore over-winter in their rookeries. Actually, the egg-laying season at a rookery of kings extends over a period of some five months from late November until mid-April when food becomes scarce. In practice, however, the effective laying season may be said to cover only three months, for eggs laid after February and hatched after the middle of April, are unlikely to produce successful chicks, because these will probably die of exposure or be trampled on or suffocated when attempting to shelter in the crèches among the older chicks during May blizzards. Even if they survive these hazards they will probably not have accumulated sufficient reserves of fat to carry them through the winter, and will starve to death during the late winter months when the parents' visits to the rookery become infrequent. As a result of this staggered laying season the oldest king chicks are approaching full size when the youngest are just hatching.

Since king rookeries are normally situated near open water, the adults are not confronted by the emperors' problem of having to trek long distances over the ice in order to obtain food for their chicks. Moreover the latter hatch when krill are swarming, and therefore perhaps also the squid on which the kings feed almost exclusively in such localities as South Georgia. During these early months the chicks are reported to receive 2 pounds of squid or fish every hour of the day. Whether or not this is actually the case, the feeding of young penguins is certainly concentrated, for the weight of an Adélie chick is increased 16-fold in its first 12 days, from 3 ounces to 3 pounds! But parental care among the king penguins varies from one individual to another. Some pairs desert their chicks before they have begun to molt at midsummer. Others feed theirs persistently until they are almost fully feathered—that is to say for a period of from 10 to 13 months through the winter until the following midsummer.

Thus the breeding cycle (from the laying of the egg to the departure of the young) may extend over a period of from 14 to 18 months, compared with the 8 or 9 months of emperors and less than 6 months of small penguins, and parental care can, and normally does, continue from one season into the next. Those parents who feed their chicks most frequently and efficiently, and are therefore the first to free themselves of parental ties and go into molt, complete the cycle in 14 or 15 months. These can therefore, in theory, rear young in successive years and breed twice in three years: whereas those taking more than 15 months must, like wandering albatrosses, miss a year.

For all but the first 7 or 8 weeks after birth the chicks pass much of their time in crèches, which are the natural outcome of their initial gathering together in loose groups of twenty or thirty when alarmed or chilled by cold winds. Fed constantly by their parents, they weigh 25 or 30 pounds by the onset of winter, much of this weight being contributed by a layer of subcutaneous fat upwards of three-quarters of an inch thick. But as autumn draws on and the ice closes in, so the parents find it increasingly difficult to obtain food, and during the period June to September their visits to the rookery are intermittent. Small chicks, less than 15 or 20 pounds in weight, are deserted, while larger chicks in crèches receive food only at intervals of from 2 to 4 weeks. Individual parents may indeed remain at sea for as long as 6 or 7 weeks, while their mates return to feed the chicks and stay ashore for 2 or 4 weeks. Under these conditions the young kings are obliged to draw heavily on their reserves of fat; but nevertheless lose weight steadily between feeding visits and, like the young wandering (and also royal) albatrosses, almost stop growing during this four-month period. By the end of the winter their weight has been reduced by one-third, and those falling below 6 or 8 pounds are unlikely to survive until the spring when the adults are able to bring food more frequently.

Thus, only less extraordinary than the hatching of the young emperors at midwinter is this over-wintering of the young kings in huge crèches of several thousand, "attended" by a few adults. Clearly, their crèches serve the same purpose as the testudos of the adult emperors—namely as vital sources of mass shelter and heat conservation during blizzards and sub-zero temperatures. During the storms and darkness at midwinter they will also serve as focal points for adults returning with food, whereas if the chicks were scattered over a large area of rookery they might become dispersed and lost. Even during the spring and summer, storms can result in heavy mortality among the crèches of young emperors, despite the fact that in such conditions the latter pack into tight testudos and are surrounded, fortuitously or deliberately, by a protective rampart of adults. The heaviest losses result from chicks being blown bodily to great distances during blizzards. More than 300 of the Pointe Géologie emperor chicks were killed in this manner during the summer of 1952, in addition to another 200 buried by the collapse of a huge ice-cliff and, indeed, a channel near the rookery was named the Valley of the Innocents by the members of an earlier French expedition, because of the countless numbers of young emperors that had been swept into it over the years by the hurricane-force winds which funneled through it during blizzards.

These are not the only climatic hazards that must take a much greater toll of young penguins than predators do.

Blizzards or heavy falls of snow can freeze young Adélies to death or affix them to the ground so that they are unable to raise themselves to take food by plunging their heads into the old birds' throats. Other youngsters must succumb to "balling-up" with frozen snow. When at the emperor rookery on Haswell Island in the middle of March, Sir Douglas Mawson noted that a dozen adolescents, which had not been able to complete their molt, were encumbered with hard cakes of snow covering their eyes or by icicles dangling from their bodies, and some of the larger ones were so heavily encased in ice, especially about the head, as to be helpless. Again, in the middle of April the following year, the one remaining adolescent, apparently still in the process of molting, was on a ledge overhanging an icy cove, partially exposed to a 65-mile per hour gale with the temperature at −10 degrees F, and was smothered with snow and small icicles. The wintering adult emperors presumably create sufficient heat in their testudos to avoid this danger. Even in the spring, however, Edward Wilson observed that those files of adults returning full-fed to the Cape Crozier rookery, from leads of open water some two miles distant, were covered with ice-crystals, and that when they were about a hundred yards from the rookery they would stop to preen themselves very thoroughly, ridding their feathers of rime and crystals. Between 50 and 90 per cent of the total losses among young emperors are attributable to the hazards of blizzards, ice-falls and crevasses, and casualties are particularly heavy among those that stray away from the shelter of parents or crèches. As late as November one year the adults at Pointe Géologie were leading the chicks away from the rookery to shelter from a blizzard in a hummocky zone at the foot of a glacier. Heavy losses must result when brooding adults are "stampeded" by falls of ice from cliffs overhanging a rookery. The Cape Crozier rookery, being situated at the base of the ice-shelf, is evidently particularly vulnerable to cliff falls, and also to unseasonable break-ups of the rookery ice. The rather small numbers of adults there have always fluctuated, and very few young ones have been in evidence. When Lt. Skelton's party were ultimately able to get down to the rookery, after being blizzard-bound in their tents for six days, they could locate only about 30 living chicks (and some 80 dead ones) with about 400 adults. A fortnight later, when several hundred adults were present, Edward Wilson could not find any young, and those seen previously had evidently died. In September the following year Wilson estimated that more than three-quarters of the chicks had been lost, either as a direct result of an ice-fall or of the break-up of the rookery ice, or because these catastrophes had caused their parents to desert them. Later that spring Wilson was blizzard-bound for ten days out of the three weeks he was able to spend at the Cape. On the afternoon before the storm broke he and his companions were standing on an old outlying cone of Mt. Terror, some 1,300 feet above the rookery on the bay ice, with the Ross Sea completely frozen over—a plain of firm white ice as far as the horizon. No crack of open water could be seen, although normally a lead ran all the way along the ice-shelf. Nevertheless, despite this appearance of stability in the ice, the emperors were unsettled—affected perhaps by the falling barometer or aware because of the absence of that lead of open water along the ice-shelf, that the ice was drifting out of Ross Sea. Although the ice had not visibly begun to move, a long file of emperors was migrating out of the rookery bay to join a group of one or two hundred, who had already collected at the edge of a frozen crack about two miles out.

After watching this exodus for an hour or more, the party returned to camp. The next morning they awoke to a southerly gale with a smother of drifting snow, and the storm continued without intermission all day and night until the following morning, when it abated sufficiently for them to visit the cliffs above the rookery. A catastrophic change had taken place. For a distance of 30 miles, out to a long line of white pack-ice just visible on the horizon, Ross Sea was open water, though large sheets of ice were still drifting north. The emperors were again on the move with a file of a hundred or more walking out to join two companies waiting at the edge of the now open sea on a projecting piece of ice—the next that would break away. The line of tracks in the snow, left by those who had migrated earlier, were cut off short at the edge of the water, indicating that they had gone out to sea, while the numbers of those remaining in the rookery under the ice-cliffs had been reduced by almost half in six days. Two days later the migration was still in progress, though it apparently comprised only those emperors—the vast majority—who were not brooding chicks for these were huddled under the ice-cliff, sheltering as best they could from the storm. By October 28, when the exodus was in its eleventh day, no ice could be seen in Ross Sea, the rookery ice was gradually being eaten away, and only a remnant of the original colony remained.

It is possible that in some years none of the Cape Crozier chicks survive to fledge and go to sea. On the other hand at both Haswell Island and Pointe Géologie between 70 and 80 per cent of those that hatched in observation years did survive, and it is significant that the rookeries at both Cape

Crozier and Haswell Island appear to have retained stable populations throughout this century, though their numbers have fluctuated from year to year. Indeed the Crozier population appears to be the same now as it was seventy years ago—between 1,500 and 2,000 pairs. A high mortality rate may be the rule, rather than the exception, among the birds of the Antarctic. Wilson's petrel, for example, has suffered a regular 65 per cent mortality at one station; but is, nevertheless, one of the most numerous of all birds. There is no evidence that the emperor penguin is a declining species in imminent danger of extinction. On the contrary, it is probably a successful one.

9: Penguins and Their Enemies

We left the young emperors in their spring crèches. They have reached weights of 25 pounds, and are consuming as much as 8 or 10 pounds of food at a single meal, composed mainly of fish with some krill finely ground into a paste from which shells and skeletons have been removed. On this extremely rich diet their weight continues to increase rapidly to 35 or 40 pounds, and in the early summer their parents cease feeding them. At some rookeries, such as Haswell Island, the latter are still having to trek 20 miles over the ice to the open sea; and now, on returning from fishing, they ignore the young ones and settle down to sleep. By contrast, the young kings (twice as old as the young emperors), though breaking away from the crèches in the summer to paddle independently in the coastal shallows off the islands, still pester the adults for food. By November they have increased their weight to 40 pounds or more, after their semi-fast during the winter, and are in good shape to withstand the debilitating molt, before leaving for sea in their second autumn.

Meanwhile, during the very last days of November and the first fortnight in December at the Pointe Géologie rookery, and some weeks later at more southerly rookeries, three stages of the emperors' life cycle coincide—the young ones prepare to depart for sea, with the majority leaving during the second and third weeks in December; the adults begin to molt; and the previous year's adolescents return to their rookeries for the same purpose, presumably because they are unable to swim among the broken pack-ice while molting, since for a period of from two to four weeks both the waterproofing and insulation of their plumage is ineffective, and immersion leaves them sodden and chilled. For the chicks this is another critical period. Though they can in fact swim short distances when still in down, they obviously cannot withstand prolonged immersion until they have completed their molt into juvenile plumage, in which they are fully waterproofed by the "chainmail" of feathers that have been growing beneath the down. This molt must therefore, preferably if not essentially, be accomplished before the rookery ice breaks up. If all goes according to plan the chicks, still only three-quarters grown, congregate in groups at the edge of the ice and wait for it to disintegrate and provide them with floes on which they can raft away to sea. In 1952, for example, the break-up of both old and new ice was almost complete at Pointe Géologie by the end of December. Open water was lapping the edge of the rookery ice, now greenish in color and hollowed in places where testudos had formed. And large floes were drifting out to sea with crowds of imperturbable emperors, beaks upraised, flippers held stiffly to sides, standing on their "decks." Nevertheless, at both Cape Crozier and Haswell Island the young emperors are to be seen migrating out to the edge of the ice when still in full down in November and December, and it seems inevitable that in some years they will have to complete their molt after they have gone out to sea on the floes. On November 25, for example, Sir Douglas Mawson counted some 7,500 emperors, the great majority of them young, covering 4 or 5 acres of floe-ice and the lower slopes of several bergs about one mile off Haswell Island; and Scott's ship, the *Terra Nova*, encountered a solitary young one in the pack-ice 750 miles off Cape Crozier as early as December 19. Adults in various stages of molt are also encountered in the pack. On January 8, for example, when 450 miles from the Ross Ice-shelf, the *Terra Nova* sighted three adults, one in molt, on a low hummocky berg. The ledge on which they stood was too high above the water-line for them to have regained it if they had once left it, and its soiled surface indicated that it had been occupied for some weeks. We do not know where the young emperors pass their first year at sea; but they presumably do so in the pack. Emperors are very rarely sighted north of the Polar Front, though a few have been reported at such cold-temperate islands as the Falklands, and one, early in April, on the south coast of New Zealand's South Island.

The majority of the adults and two-year-olds complete their molt at Pointe Géologie towards the end of December, and the rookery breaks up after nine months' occupation. Only a few sick adults and a score or two of weakling chicks, with virtually no fat on them, remain. One finds it difficult to credit that the adults, after so long and arduous a season, can be in breeding condition again within a bare three months of completing their molt. They must therefore breed in alternate years and, if that is the case, the estimated populations of the various rookeries must be doubled.

The first few weeks or months of the young emperors' life at sea, before they have learned to swim efficiently and evade the attacks of leopard seals and killer whales, may be the most hazardous of their adolescence. Although one assumes that killer whales take a considerable toll of penguins I have in fact been able to trace only one eyewitness account of them doing so—in Ellery Anderson's *Expedition South.* He noticed one morning that the Hope Bay Adélies were "porpoising" at high speed over the water and popping up on the ice-foot in hundreds, while the floes in the bay were black with crab-eater seals. A few minutes later five killers surfaced with snapping teeth as they slaughtered the nearest Adélies. We do not know to what extent the emperors are preyed upon by leopard seals at sea. Inshore, although there may be as many as a dozen dispersed over McMurdo Sound in the early summer, intercepting the adult emperors on their fishing expeditions to and from Cape Crozier, the young emperors happily leave for sea at that season when the leopards are least numerous around Antarctica. It is the penguins of the Antarctic islands which are their main prey; and Murray Levick (Scott's surgeon) shot one leopard whose stomach contained no fewer than eighteen Adélies in various stages of digestion, while its intestines were stuffed with feathers remaining from the disintegration of many more. In view of the leopards' agility on the ice one wonders why they never prey on the occupants of those rookeries situated at no great distance from the sea.

The young kings, like the young emperors, benefit from the fact that their marine life begins in the late summer or autumn when leopard seals are least numerous in coastal waters; but almost every day from May until early in November one or two leopard seals are to be seen swimming to and fro 50 or 100 yards off the rookery beaches of South Georgia, apparently capturing as many adult kings and subsequently gentoos as they require from the constant stream of birds returning from fishing. However, when engaged merely in refreshing themselves in the waters adjacent to a

Leopard seal

rookery, the kings appear to have developed measures of defence against attacks by leopards. Bernard Stonehouse observed that they are especially alert when intending to bathe, refraining from entering the water if a seal or any object resembling a seal is visible. Moreover, once they are in the water, groups are liable to frequent, sudden panics, which result in all the birds rushing out of the sea and up the beach. These panics are apparently induced by a member of the group slapping the water with its flippers and producing a sound resembling that made by a paddle striking the sea. And Stonehouse noticed that on the few occasions when a leopard seal had been able to approach a group of kings undetected, it was invariably frustrated by one of these curiously fortuitous panics, which drove the kings out of the sea before it could make a capture. Those kings departing for extended fishing expeditions take the opportunity to do so during the daily morning bathing parades, forming groups of their own and swimming in wider and wider circles, calling excitedly, before ultimately porpoising out to sea at high speed. During the summer when leopards are rarely present off South Georgia, there are no panics.

Despite a leopard's undoubted agility and speed, as it circles, confuses and exhausts its victim, and its ability to leap after a penguin on to a floe 8 feet above the water, hunting penguins successfully may not be all that easy. Streamlined for speed, wings hard and firm for beating against water instead of air, chest muscles and keel heavy and strong —an emperors' pectoral muscles measure 4 inches—a penguin can perhaps accelerate in evasive underwater bursts to speeds approaching 30 knots. Even when swimming by porpoising on long fishing expeditions, shooting out of the sea and projecting itself through the air in graceful curves in order to refill its lungs without interrupting its forward progression, it can maintain a speed of 12 to 15 knots. The leopards may therefore have to resort to stealth in certain circumstances. In December, for example, when the Adélie chicks are in process of hatching and only the adults are frequenting the sea, much floe-ice may drift in to South Georgia and then drift out again two or three miles as winds and currents shift. Although no beaches are exposed at this season, the ice over-hang all along the coast constitutes a hazard to any Adélies entering the water, because it conceals the waiting leopards; though, on the other hand, it benefits those penguins returning from fishing, since they can approach the shore at maximum speed and leap spectacularly on to its surface, as much as 12 feet above the water.

The Adélies are patently aware that leopard seals are likely to be waiting for them in the sea, and there are prolonged periods of hesitation among those preparing to go fishing, before they finally plunge off the ice-foot. This they may do all together—a stratagem which either confuses the leopard seal or enables all the Adélies to escape except one unfortunate. But, more usually, they remain lined up at the edge of the ice until, in their nervous jostling, one of their number ultimately overbalances and falls into the sea. Whether, however, its companions then wait to observe what happens to the "guinea-pig" is another matter, for there may have been confusion here with a game that Adélies play when leopards are not about. In this they crowd up to the edge of the ice and endeavor to push one another into the sea. If one is pushed in he turns quickly underwater and bounds on to the ice again. This game may last for a few minutes or for as long as an hour, but ultimately one Adélie dives in voluntarily, and is immediately followed by the remainder.

In January the Adélies are still going out to sea in large parties, but the conditions are now different. The ice overhang has melted, the beaches of black sand are exposed, and the waiting leopard seals are now stationed just outside the zone of breakers. Those penguins returning to the rookery have now to contend with the surfing breakers, and some are buffeted back and forth between the blocks of ice fouling the shore, affording a leopard the opportunity to knife through the surf and seize a penguin before it can make its dash up the beach; while the young Adélies, flailing ineffectually with their flippers on their first venture into the sea, are easy prey. In these conditions the American naturalist, Robert Cushman Murphy, saw one leopard catch seven Adélies, all but one of them young, in the space of seventy minutes. Some of these were, however, freed after several minutes, unharmed except for a few small tooth punctures; for even when satiated the hunter cannot forego the excitement of the chase.

It was Murphy who first drew attention to the peculiar structure of a leopard seal's windpipe, which, instead of being a tube as in the case of other mammals, is a flat band, opening only when the seal breathes. This flattening allows exceptional space for the gullet and enables the leopard to swallow a penguin almost whole. If the victim is one of the small penguins the leopard thrashes it on the water and shakes it, as a dog does a rat, while tearing it apart or in order to skin it, and it often bites off the head and feet before swallowing the carcass. But the leopard seal does not separate skin from body with one flick of the head, as is so often asserted, and the faeces of many leopards are composed almost entirely of penguin feathers. If the victim is the larger king or emperor, the leopard may whip it back and forth until skin and feathers peel away and slip up round the penguin's neck; the large breast muscles and viscera are then eaten, and the remainder of the carcass abandoned to the skuas after a quarter of an hour or so.

Not all penguins are ultimately killed by predators or perish in blizzards or in storms in the pack-ice. Man has taken and continues to take his toll of penguins, as he does of deplorable numbers of seals, to provide food for his sledge-dogs which even in this mechanized age are still essential to polar expeditions. When man concludes an expedition, buildings and stores abandoned in the area of a rookery result in the formation of deep drifts of snow whereas previously the ground had been swept bare by the winds. With the thaw the drifts melt into pools which inundate the nesting penguins. Emperors, even in the remotest rookeries, lose eggs and chicks when frightened by motor sledges and low-flying planes, no matter how punctilious the expedition's personnel. Adélies, in particular, have suffered heavy losses when immature birds, prospecting for permanent nesting sites, have been frightened away at the critical period by helicopters or by trigger-happy photographers; and also when helicopters have scared adults off chicks or eggs and created the opportunity for skuas, flying close to the ground in order to avoid the down-wash of air from the rotor-blades, to swoop on them.

It is possible that man actually introduces diseases into isolated penguin communities, as he has so often introduced them into isolated Arctic communities of Eskimos and Indians. At Pointe Géologie, for example, large numbers of young emperors have been affected by a paralysis, which Jean Rivolier feared might be an encephalo-myelitis virus transmitted by human agency, and which was introduced in this instance by the members of Marret's expedition when handling the emperors during the course of their study. Again, although the numerous penguin populations of the Antarctic islands appear to have recovered from their decimation by whalers and sealers in the nineteenth and early twentieth centuries, the populations of other birds have been fundamentally affected by man's whaling activities, and especially by the establishment of shore stations at which the whales were butchered and processed. (In 1965 this operation was transferred to factory ships, and the shore stations abandoned.) On South Georgia, in particular, the station offal provided an immense seasonal surplus of food, attracting unnatural breeding concentrations of giant petrels, Cape pigeons (pintado petrels) and Dominican gulls. Normally Cape pigeons, whose main breeding grounds are the South Shetlands and Orkneys, feed on krill and other plankton; but at South Georgia they fed almost exclusively on the oily globules and small blobs of fat that flowed off the flensing plants, and flocked in thousands to every whale carcass. But when the shore stations closed down in the winter the giant petrels, squatting on the beaches below the snow-foot, emaciated, stomachs empty except for a few blades of tussac, waited for death—although the immatures of their kind were winging over the southern oceans as far afield as Australia, New Zealand and the Pacific Islands—and thousands of other petrels and gulls, instead of emigrating, also remained to starve on the island, though some gulls scavenged throughout the winter on the carcasses of young kings. It is not the uniquely harsh environment but man who, directly or indirectly, threatens the survival of animal life in the Antarctic.

Some penguins, however, apparently survive every hazard of climate and predator and die of genuine old age, when perhaps twenty or thirty years old, if emperors or kings. Neill Rankin has described in *Antarctic Isle* how, when visiting a colony of kings on South Georgia in 1947 he found that:

> Well-worn paths led up hill from this colony to higher slopes where the snow was fresh and clean. . . . Following one of these I came upon a small pool in a fold of the hill-side, round the edge of which a number of penguins were standing, all facing in towards the center. I wondered whether I had rediscovered a curious place which Dr. Murphy had found in 1913.

In *Logbook for Grace*—strangely, a series of enchanting loveletters—Murphy describes how on January 10, 1913 he visited two colonies of gentoos on South Georgia and noted that:

> We have often remarked upon the extraordinarily few dead penguins encountered among the large populations. Now I have discovered their romantic sepulchre.
>
> Near the summit of a coastal hill I came upon a lonely pond in a hollow of ice-cracked stones. Several sick and drooping penguins were standing at the edge of this pool of snow water, which was ten or twelve feet deep. Then, with a tingling of my spine I perceived that the bottom was strewn, layer upon layer, with the bodies of Gentoo penguins that had outlived the perils of the sea to accomplish the rare feat among wild animals of dying a natural death. By hundreds, possibly by thousands, they lay all over the bed of the cold tarn, flippers outstretched and breasts reflecting blurred gleams of white. Safe at last from sea leopards in the ocean and from skuas ashore, they took their endless rest; for decades, perhaps for centuries, the slumberers would undergo no change in their frigid tomb.

This extraordinary observation, of which one might have doubted the authenticity had the narrator not been Murphy, perhaps explains the presence of those mummified Adélies in the dry oases of Antarctica.

10: Links

Amazingly, there may be living links between Antarctic and Arctic, despite the fact that these two polar regions are some 12,000 miles apart. Arctic terns, whose breeding range stretches as far north as there is land and as far south as the British Isles and the Gulf of Maine, winter in the South Atlantic and Antarctic. In the latter, as in the Arctic, they frequent the pack-ice, feeding on krill in the leads and in the vicinity of icebergs. A phenomenon of the distribution of migrant species is that the further north a bird breeds the further south it migrates to winter quarters, overlapping those of its kind that breed less far north and winter correspondingly less far south. Thus, it is a fair possibility that those Arctic terns reported "wintering" off Atlantic-facing coasts of Antarctica have in fact summered in the Arctic, rather than on the coasts of Scotland or Massachusetts. At one time it was thought that Wilson's petrel, breeding as far south as Antarctica, performed this epic migration in reverse. But though these small petrels, only half the size of Arctic terns, definitely range 7,000 miles north to the Newfoundland Banks, there do not in fact appear to be any satisfactory records of them in Arctic waters. The Arctic tern is virtually indistinguishable from the wreathed tern of Antarctica and the Kerguelen tern, which disperse no further north than the southern oceans during the winter. Where did the common ancestor of these three species originate? A similar problem is posed by the silver-gray petrels, which also range no further north than the southern oceans from their breeding places on Antarctica and the islands, and their counterparts, the fulmar petrels of Arctic and temperate latitudes, which winter no further south than the north Atlantic, since neither traverse the tropics.

But terns are not the only birds of the northern hemisphere to visit the Antarctic, for in January 1951 a first-winter Arctic skua—whose breeding range extends from the Scottish Highlands and Islands to within 10 degrees of the North Pole—was shot on Signy Island in the South Orkneys. It was associating with the local great skuas which are widely distributed throughout the Antarctic, with various races, distinguished by relative size and depth of coloring, on Antarctica and all the island groups. They too have their counterparts in the bonxies of the northern hemisphere, though in this case it is possible that wanderers from south and north overlap in the central Atlantic. There is an extraordinary similarity between these great skuas of the two hemispheres, both in appearance and behavior. Again, there must have been a common ancestor but, whatever their place of origin, those settling in the polar habitat of the Antarctic have apparently been much more successful than the non-polar bonxies. The latter seem never to have been numerous, and their total population in Iceland, Faroe and the north and north-west islands of Scotland is currently estimated to number no more than about 8,000 pairs, some 5,000 of which are based in Iceland.

There are, however, no penguins in the northern hemisphere, though environmental conditions are apparently not unfavorable. Some thirty-five years ago, for example, a few macaroni and king penguins, both of which are almost

Wilson's petrel

restricted to the Antarctic, together with jackass penguins from South African waters, were transported to the Lofoten Islands off north Norway, and though none are known to have bred, some survived in these northern waters for at least eighteen years.

The penguins' potential niche in the northern hemisphere is occupied by the totally different auks. Until a century and a quarter ago the latter included the great auk which, like the penguins, was flightless. The remarkable fact about this auk was that, though flightless, it was never in any danger of extermination by natural predators. Although it nested at low levels on the rock shelves and platforms of stacks and islands, these were outwith the normal range of polar bears and were too far from the mainland to be accessible to foxes. The history of the great auks is one of continual massacres by man, culminating on a day in the first week of June 1844, when the last two ever to be seen alive were killed on the flat-topped stack of Eldey, 14 miles off the south-west coast of Iceland. Now, the outstanding feature of the penguins' geographical distribution is that there are obviously only a limited number of regions in which a flightless bird, slow-moving on land, can breed in immense concentrations and rear its young without suffering an intolerable mortality from terrestrial predators. Such sanctuaries are to be found predominantly in the southern hemisphere, where there are no polar bears, wolves or foxes. Actually of the 18 species and sub-species of penguins, 4 breed on sub-tropical islands, with the Galapagos penguin able to live almost on the equator because of the wealth of food in the cool Humboldt current; 7 species breed in south-temperate seas; 5 in regions washed by cold northing currents; while only 2, the emperor and the Adélie, are restricted to the Antarctic. In the water great auks were perhaps preyed upon by sharks and possibly by killer whales, though not to the degree that Antarctic penguins are by leopard seals; but both found it more profitable to employ their wings as swimming-flippers with which to hunt fish, squid and krill, than for flight, despite the fact that their inability to fly obliged some species of penguins to climb or walk scores of miles over rock, tussac or ice between feeding waters and breeding rookeries. In this respect, when the Antarctic diving petrels (which are perhaps the penguins' closest relatives) molt they shed all their wing feathers at one time, and during this flightless period must obtain their krill and fish by using their wing-arms as flippers and "flying" underwater. The smaller species of auks, on the other hand, found it more profitable to retain their powers of flight in order to fly up to nesting sites on cliffs hundreds of feet high, and to procure their food by diving.

Possibly the ancestors of penguins originated in regions south of the equator; but the warm equatorial seas now present an impassable barrier to their flightless descendants, whose range reaches its northern limits in the cool Humboldt and Benguela currents. So long as adequate supplies of food were available within tolerable distances of breeding rookeries, climatic conditions of extreme cold would be comparatively unimportant because birds, being warm-blooded with a relatively constant body temperature, are equipped with their own heating system which renders them independent of their environment and enables the wintering emperors to survive lower land temperatures than any other animal.

II
THE ARCTIC

Tell me, Father, what is the white man's heaven?
Is it like the land of the little trees when the ice
has left the lake? Are the great muskoxen there?
Are the hills covered with flowers? There, will I
see the caribou everywhere I look? Are the lakes
blue with the sky of summer? Is every net full of
great, fat white-fish? Is there room for me in
this land, like our land, the Barrens? Can I camp
anywhere and not find that someone else has camped?
Can I feel the wind and be like the wind? Father,
if your Heaven is not like all these, leave me alone
in my land, the land of the little sticks.

DOGRIB INDIAN TO OBLATE PRIEST

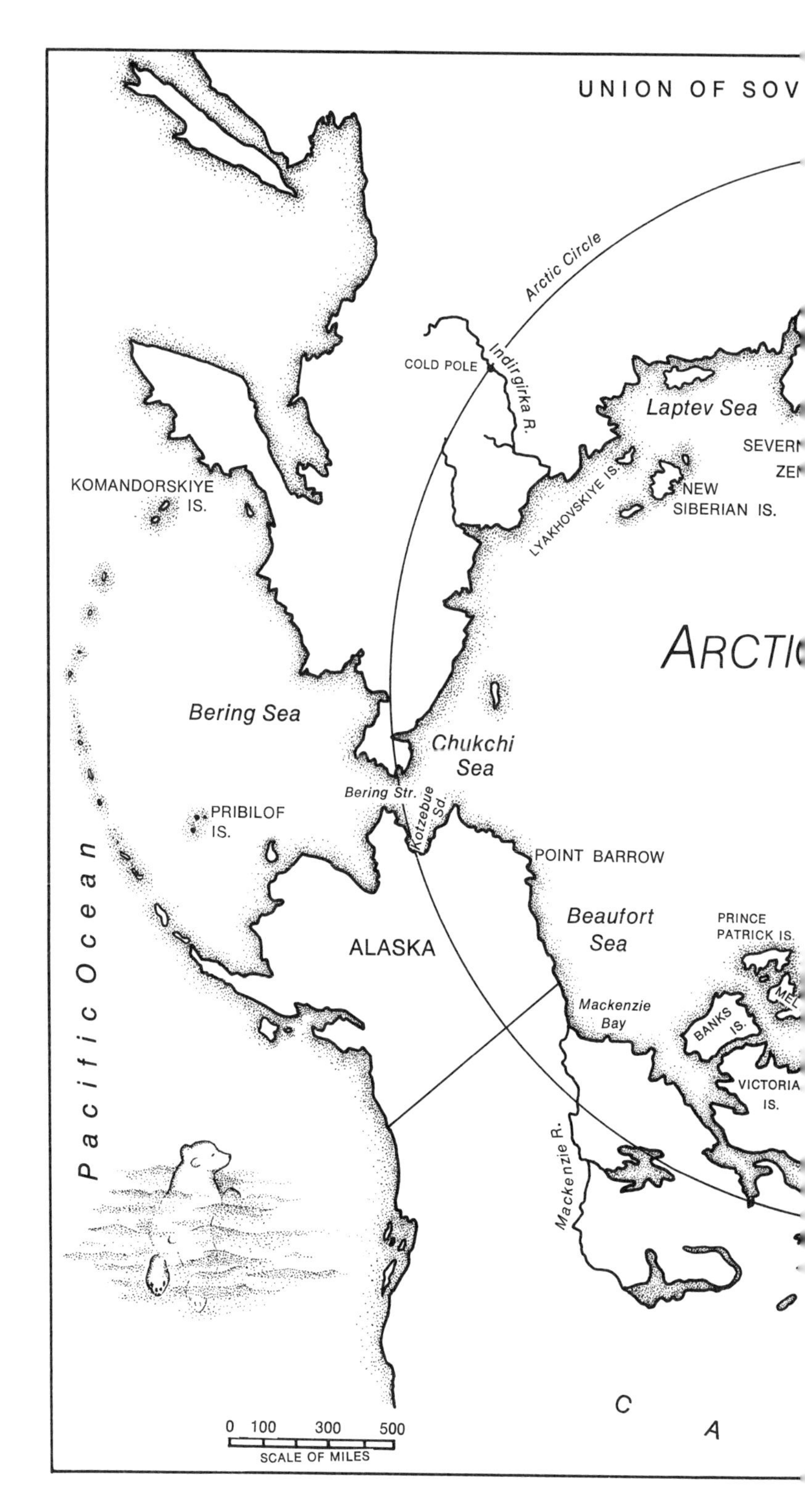
UNION OF SOV
Arctic Circle
COLD POLE
Indirgirka R.
Laptev Sea
SEVERN
ZE
LYAKHOVSKIYE IS.
NEW
SIBERIAN IS.
KOMANDORSKIYE
IS.
ARCTI
Bering Sea
Chukchi
Sea
Bering Str.
Kotzebue
Sd.
PRIBILOF
IS.
POINT BARROW
Pacific Ocean
Beaufort
Sea
PRINCE
PATRICK IS.
ALASKA
Mackenzie
Bay
BANKS
IS.
VICTORIA
IS.
Mackenzie R.
C
A
0 100 300 500
SCALE OF MILES

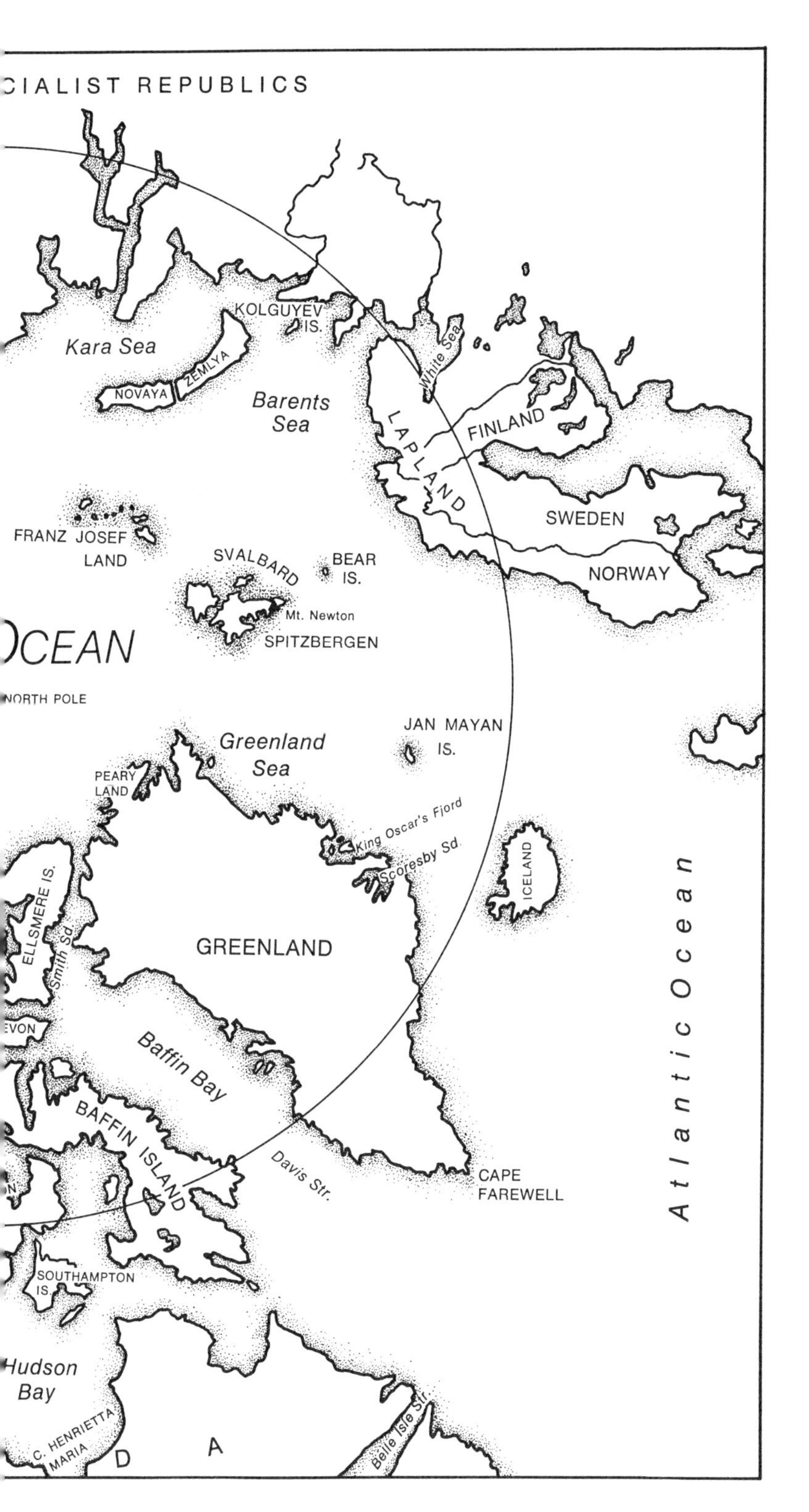

CIALIST REPUBLICS
Kara Sea
KOLGUYEV IS.
NOVAYA ZEMLYA
Barents Sea
White Sea
LAPLAND
FINLAND
SWEDEN
NORWAY
FRANZ JOSEF LAND
SVALBARD
BEAR IS.
Mt. Newton
SPITZBERGEN
OCEAN
NORTH POLE
JAN MAYAN IS.
Greenland Sea
PEARY LAND
King Oscar's Fjord
Scoresby Sd.
ICELAND
ELLSMERE IS.
Smith Sd.
GREENLAND
EVON
Baffin Bay
BAFFIN ISLAND
Davis Str.
CAPE FAREWELL
Atlantic Ocean
SOUTHAMPTON IS.
Hudson Bay
C. HENRIETTA MARIA
D A
Belle Isle Str.

11: The Nature of the Arctic

In contrast to the Antarctic, which comprises a frozen continent surrounded by a partially frozen sea, the Arctic is a partially frozen ocean almost surrounded by land. Since both sea and land are embraced by the same climate how is a naturalist to determine the boundaries of his Arctic? Clearly, these cannot be demarcated by an oceanic Polar Front. But just as the latter is associated with the 43-degree F isotherm, so the naturalist's Arctic can be satisfactorily defined as including those regions wherein the mean temperature for the warmest month of the year does not reach 50 degrees F, nor that for the coldest exceed 32 degrees F. Since trees cannot normally grow where the July temperature does not reach 50 degrees F, this isotherm marks the northern limit of most forest growth, though blocks of stunted conifers have been able to establish themselves along the northern coasts of Russia and in the lower Lena valley. But, by and large, tundra and ice take over from the illimitable coniferous forests of the taiga along the contours of the 50-degree F isotherm. Conveniently, its contours also delineate the northern limit reached by many marine animals inhabiting temperate seas, and the southern limit of others inhabiting Arctic waters. The naturalist's Arctic therefore includes not only the Arctic ocean and its islands but such land areas as northern Siberia, the northern part of Iceland, the whole of Greenland, a million square miles of North America from

Labrador north-west to Alaska, and the Canadian Arctic islands.

According to marked climatic differences, reflected in the distribution of plants and animals, this polar region can be divided into High Arctic and Low Arctic. Even in the former, which includes such land masses as the Canadian Arctic islands, north Greenland and the Siberian islands, the climate is markedly less severe than that of Antarctica. A brief, very cool summer with a July mean below 41 degrees F is followed by a stormy winter with temperatures almost permanently below freezing. Nevertheless, the mean annual temperature on the 10,500-foot crystal dome of the Greenland Ice-cap, which blankets the whole of the island except for a coastal strip nowhere wider than 200 miles, is estimated to be no lower than −27 degrees F, and that at the North Pole no lower than −9 degrees F. Away from the inland plateaus temperatures rarely fall below −70 degrees F over the land or −50 degrees F over the sea, though the winter concentrations of ice-floes, up to 13 feet thick, do not begin to break up until July or August. The Cold Pole in the northern hemisphere is far inland from the moderating influences of the Arctic ocean's mass of water, and is located on the upper Indigirka in the southern highlands of eastern Siberia, where the thermometer can fall to −93 degrees F and rise to +95 degrees F in the summer. Siberia and the Yukon were believed to be the coldest regions on Earth before meteorological stations were established on Antarctica.

Climatic differences between High and Low Arctic resemble those between Antarctica and the Antarctic islands. As on the high plateaus of Antarctica, so in the High Arctic the annual snowfall is slight, ranging from only 1½ inches on some of the Canadian islands to a maximum of 20 inches.

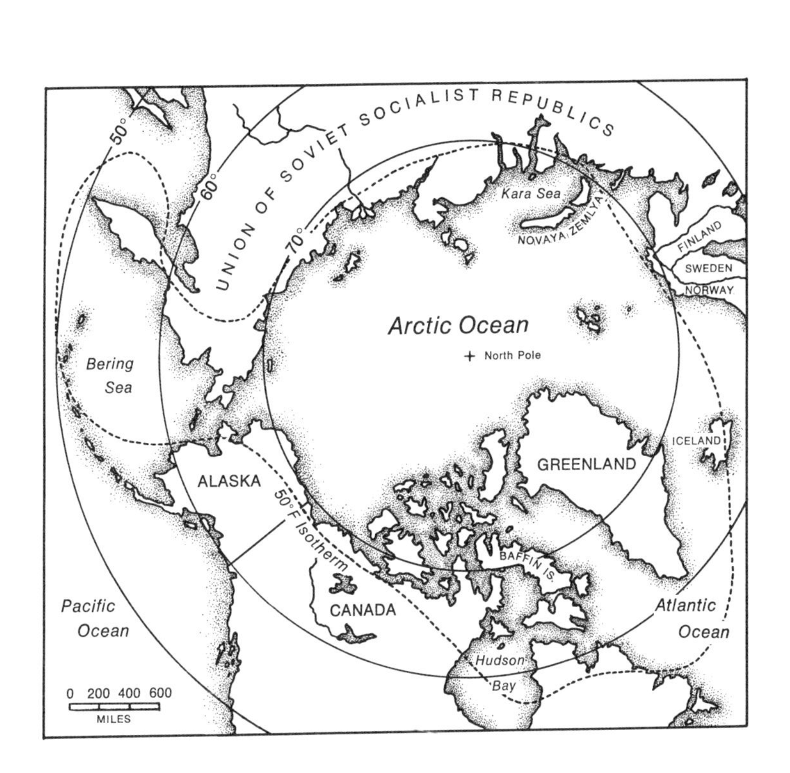

In the Low Arctic, on the other hand, although the snowfall is much heavier the winters are shorter and less severe, with freeze-up and thaw sometimes alternating, and the summers longer and milder with a July mean between 41 and 50 degrees F. Exposed coasts, in contrast to those of the High Arctic, may be free of ice throughout the winter, permitting the existence in the tidal zone of anemones, sea-urchins, lugworms, snails, mussels and various crustaceans, sensitive to the presence of elements of temperate waters which do not flow as far north as the High Arctic. This marine fauna makes it possible for purple sandpipers, feeding on mollusks, crustacea and amphipods among the seaweed on rocks and mud banks, to be the only wading birds to winter in the

Arctic. Since during the winter the majority of waders obtain their food on tidal mudflats it is evident that summer residents in the Arctic must emigrate after the breeding season to regions where shores and estuaries remain unfrozen. But why do no turnstones, which nest as far north as west Greenland and Spitzbergen, and are the inveterate companions of purple sandpipers, winter in the Arctic? Is their failure to do so associated with their unique feeding technique?—for while both species feed on the same invertebrates the turnstones obtain these by energetically pushing over fronds and mats of seaweed with their beaks and foreheads, exposing the small game beneath. But during the Arctic winter the seaweed freezes as soon as the tide recedes.

Ice-free waters off southern Greenland have also made it possible for a race of mallards to be permanently resident in the Low Arctic. Since all freshwater lakes and rivers freeze over in October the mallards are obliged to live at sea for the greater part of the year, feeding at low tide on small crustaceans and mollusks awash in the beds of seaweed, large areas of which normally remain free of ice even when the sea is frozen. As an adaptation to this marine life they, like true sea-birds, have developed very large nasal glands embedded in the skull immediately above the eyes. These not only enable them to distil fresh water, but also function as kidneys, concentrating salts and discharging a strong brine through nostrils or beak, thus disposing of any surplus saltwater imbibed while feeding. But although the mallard thrive in this arctic habitat, and are larger than those of any other race, they have reached their tolerable northern limit. They are the first Greenland birds to starve in hard weather and suffer heavy mortality in prolonged snowfalls or when the ice becomes densely packed.

As in the Antarctic, the dominant environmental factors on the Ice-cap and the islands of the High Arctic are snow over the land and ice in the sea. The marine life of the High Arctic is restricted during the winter to ringed seals, equipped to survive in the heaviest ice; a few white whales and narwhals far out at sea where the pack-ice is broken; and that remarkable auk, the black guillemot or tystie, which by diving to depths of 60 or possibly 90 feet in the tidal cracks that open off coasts and stranded icebergs, or where strong currents flow, is able to feed on euphausians among the seaweed and on the sea bottom at least as far north as latitude 82° off Greenland. Curiously enough, all tysties wintering in the High Arctic appear to be juveniles, and no species of sea-bird is 100 per cent resident. Sabine's gull, for example, which breeds fairly widely on swampy tundra shores of the High Arctic and at least as far north as the Canadian Arctic islands and Svalbard, emigrates out of the Arctic in the autumn to winter in seas as far south as the Bay of Biscay or the warm-water zone between Panama and Ecuador; while glaucous gulls retreat to such Low Arctic regions as Novaya Zemlya and southwards into the north Atlantic.

On the other hand, consider the inexplicable behavior of the beautiful Ross's or rosy gull, about which there has been little information since the Russian naturalist Buturlin first discovered its breeding grounds in 1905. Here is a bird that nests outside the Arctic, as we have defined it, in the lower reaches of three great rivers of eastern Siberia—the Indigirka, the Alexea and the Kolyma. It does not even nest on the tundra, but on mossy flats between the pools in low-lying marshy wastes of alder scrub, which may be as much as 100 miles south of the tree-line in places. To these, where it breeds in small colonies and often in association with

Arctic terns, it does not return until the last days of May or early in June, when the ice on the rivers is only just beginning to break up and snow still lies deep in the marshes. Yet some time in July, after a breeding season of only six or seven weeks, both adults and young (still with downy heads) emigrate northwards to the coast and out over the Arctic ocean. By mid-July or early August they are off the New Siberian Islands, and for a few weeks thereafter are continually on the move among the pack-ice in search of open water and the small crustaceans and plankton they pick up from the surface of the sea while in flight, although at the breeding grounds they are apparently mainly insect feeders. But late in September or early in October, when the waters over the broad continental shelf off eastern Siberia freeze solid, they are reported to assemble in flocks of hundreds or thousands to migrate eastwards or north-eastwards past Point Barrow on the polar coast of Alaska. Before the end of October all have passed on, apparently, into the northern sector of the Beaufort Sea: that ultima thule of frigid desolation, where the pressure in the pack is so intense that floes are forced up one on top of another to such heights that they block the horizon from a ship's crow's-nest 100 feet above the sea. Then, for seven months the entire population of rosy gulls, except for the odd wanderer to such outposts as the Pribilof Islands, disappears into that ice desert where conditions must approach in severity those experienced by the emperor penguins in their rookeries in the Antarctic. On what do they feed during this long immurement in the pack-ice? A Russian report states that they actually winter in open waters in the pack-ice, but although they are known to feed on the ice-shrimps that swarm around icebergs and, according to whalers, on the offal from whale carcasses, there are not likely to be many of the latter in the Beaufort Sea. What were the special circumstances that originally induced them to select such an extraordinary winter habitat? It is difficult to credit that they could not have discovered a less rigorous environment in which to winter until returning to their Siberian breeding grounds.

Comparable problems are posed by the no less beautiful ivory gulls, which are the most northerly of all birds, since they nest mainly between 75 and 85° N, and not south of 70°; and whose breeding grounds include the Canadian Arctic islands and the extreme north of Greenland, Svalbard (as the Spitzbergen archipelago together with Bear Island have been renamed by the Norwegians), Franz Josef Land and Severnaya Zemlya. It must be admitted that we can only speculate as to the nature of their winter wanderings, which do not carry them out of the Polar Basin; for though after the breeding season they migrate slowly southwards with the belts of drift-ice, they stay out on the northern fringe of the pack, never approaching land or coastal waters where the floes are too consolidated. The outer ice of the Barents Sea and the seas east and south of Greenland are probably their main winter quarters though, again, since these regions are inaccessible to shipping during the winter months, we can only speculate. Unlike the rosy gulls, they are apparently chiefly scavengers. Though nesting on boulder-strewn shores, or in some localities on cliffs, their colonies are always situated near extensive ice-fields on which seals lie out, and they pass much of their time watchful on the ice, flying up immediately the carcass of a seal is dumped on the ice for flensing. During the summer they always appear ravenous for any blubber, flesh or offal left on the ice by whalers or sealers, and are therefore probably largely dependent during the winter on scraps from the kills of polar bears—which often take only the blubber of seals, abandoning the carcasses —and also on the excrement of bears and seals. Small fish and crustaceans, washed on to the ice-floes, may provide an additional source of winter food; but even in the summer these gulls are reported to display a curious aversion to water, rarely alighting on it, never swimming in it, and only wading belly deep in the shallows. Morsels of blubber thrown into the water will not tempt them. A possible clue to this aversion may be found in the fact that immature wanderers to such non-Arctic seas as those off the British Isles bathe freely, perhaps because they have not experienced the icing up of their feathers after settling on Arctic waters during the winter. But there must be prolonged periods of winter storm when neither ivory gulls, nor rosy gulls, can feed for days at a time, and when they can only find a modicum of shelter from blizzards among the hummocks in the pressure-ice.

As in the Antarctic so in the Arctic, though to a lesser degree, most forms of life, except those of the tundras of the hinterland, are dependent on the produce of the sea for their existence. As in the Antarctic, countless myriads of planktonic crustacea provide the food base. They include those euphausian relatives of the krill which form part of the diet of ringed seals and narwhals—though not of white whales—and no doubt also of those baleen whales that summer in Arctic seas and range almost as far north as the High Arctic. But the euphausians perhaps constitute a lesser proportion of the Arctic food base than the copepods and pteropods or winged snails, especially the relatively large copepod *Calanus*. The latter is particularly numerous off glaciers and icebergs where surface waters are less saline, and in coastal waters where even in the winter the ice is subjected to pressure and breaks up early in the spring. The small polar or tom-cod and other fish feed on *Calanus*, as do the bowheads or Greenland whales, now almost extinct; though during the summer the latter apparently feed mainly on pteropods at depths of from 600 to 3,000 feet over the continental slope.

The tom-cod and the small 6-inch capelins consume enormous quantities of plankton, and are themselves a staple food of larger fish, seals, some whales, and birds. Wherever precipitous cliffs are free of ice and snow, millions of sea-birds —guillemots, little auks, puffins, fulmars, kittiwakes and larger gulls—nest on their ledges or in burrows on the slopes above. None could rear their young without the plankton and small fish they capture in leads in the ice, in zones of upwelling water off glaciers and large icebergs, or in the turbulence of currents and tide-races; while the young skuas, in their turn, are dependent for food on the kittiwakes and auks their parents can kill or pirate when returning from fishing. Seals, white whales and narwhals prey on the fish that subsist on the plankton. Polar bears prey on the seals, and their geographical distribution and seasonal migrations are closely related to those of their chief prey, the ringed seals. Although they supplement their diet of seal with carrion in the form of dead fish and the carcasses of walruses and whales, only the presence of seals makes it possible for them to inhabit the Arctic. Killer whales also prey on seals, but probably more extensively on the white whales and narwhals, breaking their spines with blows of their tail-flukes or striking them in their bellies and bursting their stomachs as, shooting up from the depths, they hurl them out of the water with the force of the impact.

Since the beginning of this century, and especially during

Little auks or dovekies

the past forty or fifty years, there has been a general rise in winter temperatures in the Arctic. They were indeed actually some thirty degrees above normal on Svalbard during the 1937–9 winters. This warming up has had a dramatic influence on marine life. Coal-fish, haddock, ling, herring and pilot whales (the ca'aing whales of the Faroe islanders) have all moved into the Arctic zone from temperate seas, as have the common starfish and a jellyfish (*Halopsis ocellata*). But most notable have been the northerly extension of the Low Arctic range of the capelins and the vast increase in the numbers of the large Greenland or fiord cod off Low Arctic coasts and particularly in the Davis Strait, which is now one of the great cod-fishing grounds. One effect of these northerly extensions in range has been to draw large herds of harp seals into the High Arctic from its southern fringes in quest of the choice capelins, which are so rich in oil that the Eskimos call them candlefish and dry and burn them as lights, if there is a surplus of food. Another effect has been to extend the High Arctic distribution of the ringed seals, following the tom-cod which, though spawning in coastal waters and fiords, travel as far north as the region of the North Pole. This movement of the seals may, in turn, have tended to extend the boundaries of the polar bears' domain northwards.

Sea-birds have also been affected by this warming up of the Arctic. Because of the consequent northerly recession of the pack-ice and its attendant plankton from Icelandic coasts some species of sea-birds can no longer obtain adequate food supplies within tolerable operational range of their breeding cliffs, and are unable to rear normal broods of young, with the result that little auks, for example, are now almost extinct in Iceland. If the recession of the pack continues, and obliges sea-birds to establish breeding colonies further and further north in order to keep in contact with the plankton and fish, they will be confronted by another aspect of the same problem; for when very late springs result in the ice failing to break up off the coasts of such northerly regions as Baffin Bay, east Greenland or Svalbard, the fish and plankton are, again, too far out at sea to be within operational range of the breeding cliffs. Sea-birds are also affected in a major way by one other climatic factor. Although all leave High Arctic breeding areas in October, not to return until the following spring, sudden January freeze-ups in the fiords and coastal seas of the Low Arctic may result in the deaths off southern Greenland of many thousands, especially guillemots, which have delayed emigrating to their main wintering area off Newfoundland.

12: Arctic Whales and Seals

The day to day routine, the feeding and breeding habits, and the migrations of all marine mammals in the Arctic are governed to greater or lesser extent by the seasonal movements of the pack-ice. Varying from thin newly formed ice to coarse old floes, 12 or 15 feet thick, the pack is ever in motion, drifting at rates varying from 2 to 8 miles a day, mainly clockwise around the Pole, though there are many counter-currents and subsidiary currents within the main drift. From the gigantic refrigerator of the Beaufort Sea the ice is driven by northerly winds into the Chukchi Sea, where some drifts through the Bering Strait into the Pacific, but the main stream flows on with the clock to pass north of the Siberian islands, Svernaya Zemlya, Franz Josef Land and Svalbard, where the warmer waters of the Gulf Stream push the ice up to 83° N. The pack then drifts directly down the east coast of Greenland, rounds Cape Farewell and flows to an annually variable distance up the west coast before joining another great mass of ice, which has drifted down from the High Arctic and eastwards through various channels of the Canadian islands to assemble mainly in Baffin Bay. The combined streams of pack-ice then flow south in a 1,000 miles procession of ice-floes to reach northern Labrador in November or December.

Marine mammals in polar regions are always vulnerable to the danger of being smothered under the ice if they are unable to find tide-cracks or broken pack-ice in which they can surface to breathe. Most must therefore migrate southwards before the winter freeze-up. The herds of from half-a-dozen to fifty and exceptionally 10,000 white whales (belugas) and narwhals, feeding and breeding mainly in the bottom waters of High Arctic bays and inlets, are driven out of these when new ice begins to form in the autumn and retreat to the Low Arctic. The milder climatic conditions have, however, made it possible for the belugas to extend their seasonal stay in the High Arctic, while the narwhals, which range to within 5° of the Pole, do not normally emigrate further south than latitude 70. In ice-free waters they pass the winter feeding on cod, Greenland halibut, rose-fish (Norway haddock), catfish and squid, all of which are commonly distributed at depths of 600 feet or more over coastal banks. It is a fair guess that they capture fish and squid with the aid of sonar, for narwhals, like dolphins, can only be caught in nets of large mesh, which cannot be detected sonically.

While both bull and cow narwhals possess teeth, which however in the case of the latter scarcely protrude from their sockets, one of the bull's canines has, as is well known, developed into a straight torqued tusk, up to 8 or 10 feet long and with a left-handed twist. In the very rare instances of bulls possessing twin tusks both show this sinistral spiral. The origin and purpose of this extraordinary cylindrical ivory tusk are still not determined. The Eskimos' assertion that the bull spears large fish with it has not been confirmed, though there is a case on record of a bull whose stomach contained, in addition to a squid, a skate 20 inches broad (almost three times the width of its captor's mouth), pierced by 7 inches of broken tusk. Normally, when hunting such fish as cod, a herd of narwhals do so in line abreast, crunching them in their blunt jaws and swallowing them whole. Nevertheless, Peter Freuchen, that extraordinary polar character, believed that the tusk was a practical feeding tool, with the tip of which the bull scratched the sea bottom in order to stir up halibut, its favorite fish, and also the flounders which lay concealed in the mud. He based this belief on three observations: the underside of the tip is slightly flattened, as an adaptation to the bull's position when working on the bottom; the tips of all tusks are worn down; and if they are broken off in a rock crack during the course of fishing operations, 4 or 5 inches of the new tip is soon abraded and plugged with mud and stones.

However, the tusk is generally considered to be a secondary sexual adornment employed in courtship, in masculine rivalry, or as a signaling device for communication. In support of this opinion is the bull's habit of basking on the surface of the sea for periods of several minutes at a time without perceptibly blowing or moving, but with tusks pointing skywards, while herds of fifteen or twenty bulls may be seen elevating their tusks, and crossing and clashing them as if fencing. It is unlikely that they fight with their tusks because, according to an unconfirmed statement by Freuchen, they are as brittle as glass, being filled almost to the tip with pulp and blood, and break very easily—though in aged bulls they begin to solidify from the tip downwards. If Freuchen is correct this would explain why the bulls do not spear fish with them. He added that even when a herd was trapped in the ice, with its members struggling desperately to obtain breathing places at a small hole, he had never witnessed a bull attempt to break the ice with its tusk or thrust it into the body of another bull.

Feeding and breeding in shallow waters, narwhals and belugas are especially suscepible to the danger of being trapped in the ice when a polynia (open pool) or a lead freezes suddenly. It is true that they can break new ice by striking it with humped backs, but if the only air-hole or polynia in an area of many square miles is surrounded by ice 6 or 8 inches thick, then hundreds of whales may congregate at such a *saugssat* or *savssat*—the Eskimo terms—jostling and struggling for breathing space. Thus, off the coast of eastern Siberia, as many as 250 belugas have been counted in a saugssat 150 yards long, 50 yards broad and 180 feet deep. When rising to breathe at 12- or 18-minute intervals, they sometimes did so in such numbers together that one would be hoisted high in the air with two-thirds of its tail and body out of the water. Early one spring Freuchen encountered a still larger herd of narwhals in a saugssat off Cape Melville, 200 of which were harpooned by Eskimos during the course of four days and nights, while as many more escaped through a narrow lead, and there is a record of 1,000 being killed in this way. Freuchen describes how the fountain of water the narwhals blew up whenever they breathed fell around the edges of the hole, freezing instantly and further restricting its size.

It is in March that the herds of belugas and narwhals, thousands strong, set out on their return migration to their breeding grounds in the High Arctic. The Eskimos call the milk-white belugas the geese of the sea, for the herds travel in V-formation, rolling high out of the smooth blue-black waters of broad leads through the pack, as they surface to breathe and sound their liquid curlew-like trills. The narwhals swim tusk to tusk and tail to tail, rising and sinking rhythmically in curving progression, with their white horns flashing out of the sea in regular cadence as the waves fling jets of spray over their bluff foreheads. These spring migrations proceed at a leisurely 5 or 6 knots because, though traveling initially far out to sea among the drifting pack, their way is eventually barred by the unbroken ice-fields of the High Arctic, where winter storms have piled immense floes one upon another in chaotic ridges and hummocks of ice resembling the ruggedest of mountains, although their peaks may be no more than 50 to 60 feet in height. In one of his many books on the Arctic, the American explorer Vilhjalmur Stefansson wrote:

> When you get down among such ice it is almost as if you were in a forest. You can see the neighboring hummocks and the sky above you, but you get no good view of your surroundings. When you climb to the top of even the highest crags of ice you get a view of all the other crags, although here and there a little ice-valley may open.

At intervals, however, the chaos subsides into more or less level fields of floe-ice that have broken off from the land-fast ice-foot and drifted out to sea. Gleaming blue-black leads of open water wind through these ice-fields. Some are mere cracks zigzagging through old floes; others rivers, a mile or even two miles broad, flowing freely for as far as the eye can follow them. Snow-laden floes and misshapen lumps of ice float like marble blocks on their dark waters. Even at the Pole the pack-ice is ever in motion, perpetually under pressure from currents and winds, with leads opening and closing and thrusting up ridges and tilted blocks of ice the size of houses. On the fringes of the ice-fields the whales must halt until the floes begin to disintegrate in June or July, feeding in the interim on bristle-worms, if fish and squid are unavailable.

Just as in the Antarctic the crab-eater and Ross seals inhabit the pack-ice and the Weddell seals the coastal ice, so in the Arctic the harp and hooded seals frequent the pack-ice and the ringed and bearded seals the inshore ice. The hooded (bladder-nosed) seals are as oceanic as the Ross seals, rarely hauling out on dry land or, for that matter, on solid ice, but preferring broken floes floating in deep water, and diving to great depths for such fish as halibut and rose-fish, or scouring the bottom for echinoderms and other invertebrates. Like the Ross seals they, too, are widely scattered over the pack and of a solitary nature. Even when, early in the spring, their two breeding populations, totaling perhaps 300,000 or 500,000, gather on the floes north of Jan Mayen and on those 50 miles off the Belle Isle Strait between Newfoundland and Labrador, they remain widely scattered in family groups of bull, cow and calf. The parents return to the sea as soon as the calf is weaned after two or three weeks, abandoning it to hunger on the floes for a further two weeks, and to lead a solitary existence for its first three or four years—or so it is reported, for the life histories of any animals inhabiting the pack-ice must necessarily be more or less conjectural.

By contrast, the harp (saddle-back or Greenland) seals are gregarious at all seasons; and only aged bulls, whom the Russians refer to as *odinetsy* or hermits, are to be found alone or in small pods. The three separate populations of harp seals congregate, like the hooded seals, in enormous rookeries to breed, and begin to emigrate south for this purpose in October and November from Baffin Bay, to haul out on fields of heavy floe-ice in the White Sea (where they number perhaps 1 million), Jan Mayen (perhaps 750,000), and the Gulf of the St. Lawrence and the Belle Isle Strait (perhaps 3 million). But unlike the hooded seals the members of a harp rookery are usually concentrated in only two or three vast assemblies, each covering between 5 and 100 square miles of ice, with from 1,000 to 6,000 seals to the square mile. The pups are born towards the end of February or early in March in the shelter of hummocky ice well away from the edges of the floes. This precaution does not, however, prevent very large numbers of the pups being crushed in the ice in those years when storms pile up the floes, despite the fact that the rookeries are not situated as far out in the pack as those of the hooded seals. No doubt others perish when swept away by waves breaking over the floes, for though they are known to be able to swim when only two or three days old, they do not normally enter the water until the soft and woolly birth coat of curly white hair has been partially or entirely molted and replaced by short gray hair when they are three or four weeks old. Although the cows desert the pups after suckling them for a fortnight or slightly longer, the latter quadruple their birth weight of 18 or 20 pounds during this period and are able to feed independently by the time they have completed the molt, sucking in the euphausians and small capelins that swarm around the floes. Early in May, after first moving across the ice-fields to their seaward edge and drifting passively southwards for a couple of weeks, the pups set out on their long migration northwards, proceeding at a rate of perhaps thirty miles a day, intermittently hauling out on floes to rest. In the meantime the adults, after assembling to molt on other ice-fields in even larger aggregations than at the breeding rookeries, have returned to the High Arctic in May and June, following the multitudinous shoals of capelins, which at this season are running far up the fiords to their spawning grounds, and which, together with crustaceans, herring and cod, will be the harps seals' food during the summer. One can only assume that it is by swimming fortuitously against the currents that the pups eventually join the adults at the summer fishing grounds, of whose existence they are of course unaware.

The only natural enemies of both harp and hooded seals are polar bears and killar whales—for man, who continues to decimate the breeding herds on the Newfoundland ice in the most barbarous manner, cannot be termed anything but unnatural. But although some polar bears drift down with the ice every year from Greenland and Baffin seas to the Newfoundland and Jan Mayen rookeries, they kill only a negligible proportion of the pups, and rarely visit the White Sea. Most hooded pups tend, in any case, to be inaccessible to bears because they are born on the floes at the seaward edge of the pack-ice, while, as we have seen, their family groups are widely scattered and not concentrated in dense

Killer whale (Orcinus orca)

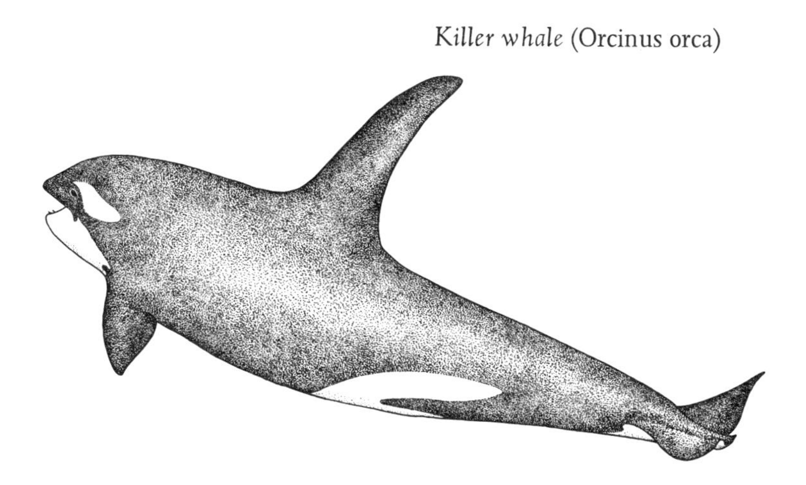

assemblies like the harp seals. For that matter, a hooded cow may exhibit all the ferocity of a leopard seal in defending her pup against a bear, and one has been known actually to kill a bear by biting through its throat, though she herself was also killed. During the summer killer whales may be the main danger to the herds of harp seals fishing among the broken pack-ice, splintering the floes with their backs and sweeping the seals off with their tails, or pressing down the side of a floe and precipitating a basking seal into the sea.

Although all seals take advantage whenever possible of cracks and leads in the ice in order to surface for air or to haul out on to a floe to sleep, the problem of finding these in the right place at the right time is with them for much of the year. The harp cows, for instance, when they first haul out at the breeding rookeries can do so through leads and holes with a relatively thin covering of ice. Saltwater ice is much more plastic than freshwater ice. A man's foot sinks through 3 inches of the former as if through a layer of tough glue, while a seal can push its nose and head up through it for a breather. The harp cows are able to keep open these holes until after the pups are born, when they will have to get up on to the ice several times a day in order to suckle them, after 15-minute fishing excursions under the ice. If the ice is not more than 3 feet thick a number of families will combine to maintain an exit hole, 2 or 3 feet in diameter, near a group of pups—presumably by constantly smashing the ice as it reforms. To be permanently resident in the High Arctic, however, a seal must be equipped with special tools for keeping open breathing holes in winter ice several feet thick. Since the large bearded seals, males of which reach 12 feet in length and 900 pounds in weight, possess only very small teeth or, in some cases, none at all, they obviously cannot maintain breathing holes in such ice; though the strong claws on their flippers enable them to scratch small holes, 2 or 3 inches in diameter, in moderately thick ice. Since some individuals have pronounced crowns of whitish scar-tissue they may, like walruses, habitually shatter thin ice with their heads. Some certainly contrive to winter under the immensely thick land-fast ice in the fiords of the Low Arctic, perhaps by sharing holes with those master ice-craftsmen, the ringed seals; and at least one bearded seal has been known to negotiate a passage through the vast wilderness of High Arctic ice, criss-crossed with black fissures, and surface in a broad lead with immense floes piled up on its edges within 2° of the North Pole. However, the majority of the 100,000 or so beardies, instead of migrating out of Arctic waters like the harp and hooded seals, ebb and flow with the drift-ice, southwards in winter, northwards in spring, and keep to the edge of the pack, particularly off Svalbard where numbers are killed by polar bears. In the summer they return to bays, fiords, river mouths and channels winding among the inshore islands, surfacing in tidal cracks and leads, alongside which the ever-watchful hunting bears tread down hard-packed trails in the snow. Only in the pack-ice, where most of them breed, may as many as fifty beardies be found in a single locality, and be heard whistling to their pups, for they are in the main non-gregarious.

Although they are reported to travel far up some Siberian rivers in pursuit of shoals of small fish, beardies are predominantly bottom feeders, except in places where the depth of water exceeds 300 feet and they must perforce hunt cod or halibut. Their bottom food includes various crustaceans, small fish such as sculpins and flounders, octopuses, rock-crabs, worms and sea-cucumbers, but especially shellfish. They do not, however, compete with walruses for the latter because their claws are not strong enough to excavate such bivalves as clams, which are the walruses' staple food. How do they obtain shellfish? Do they shovel them out of the mud with their long, extraordinarily stiff, mouth bristles or *vibrissae*, as walruses do with their very much shorter ones? Do they scrape the shellfish off the bottom with their claws and, after crushing the shells between their flippers, as walruses do, suck out the soft parts? We know that they feed on the large whelks (*Buccinium*) which are numerous on the bottom, because their stomachs may contain hundreds of those round chitinous operculums or plates, attached to the whelks' feet, with which they close their shells, and which are not dissolved by the seals' gastric juices. But the shells of other large snails, such as the 8-inch *Neptunea* and *Sipho*, are much thicker and perhaps too hard to crush.

Throughout the summer seals are able to travel freely in most Arctic waters, because every storm results in further disintegration of the fields of pack-ice and the opening up of leads through them. Even when mushy young ice forms everywhere after the first autumn frosts, a seal can still surface where it pleases because, with a sharp upward bunt of its head it can shatter new ice up to 4 inches thick and engineer a breathing hole from 6 to 18 inches in diameter. But when the ice grows thicker and also tougher, seals must either emigrate, as the harps and hooded seals do, or retreat like most of the bearded seals to the fringe of the pack-ice or, if ringed seals, settle in one locality where holes can be kept open for the duration of the freeze-up. Although the smallest of the seals, weighing no more than 200 or 250 pounds and only 4½ or 5 feet long, but with a girth equal to their length, the ringed seals have mastered the problem of ice, even to the extent of wintering in the High Arctic. The adults, indeed, seldom venture further than ten miles off the land and never more than a hundred, though if the tidal cracks and polynias in the land-fast ice freeze over, the younger seals move out to open water at the southern edge of the pack, returning to the shore ice in the spring. It is perhaps these immature seals which range as far north as the region of the Pole, if the seas there are not completely ice-bound, and into that reservoir of pack-ice, the Beaufort Sea, where Vilhjalmur Stefansson found them sufficiently numerous to supply him with food during his remarkable sledge-cum-boat expeditions from Alaska to Banks Island, and subsequently to Prince Patrick Island. With an estimated population of between 2½ and 6 million, and a life-span of upwards of forty-five years, ringed seals have clearly exploited their Arctic niche to the full.

As soon as the ice becomes permanent they begin to open up holes, preferably near cracks and in the shelter of pressure-ridges where snow piles up. Obviously, the chosen habitat of any animal must be one containing an adequate supply of food, but ringed seals are certainly strongly influenced in their choice by the condition of the land-fast ice, because the nature and area of this determine the number of suitable breeding sites. For these the seals require the most stable ice overlaid by not less than 3 feet of snow, and these conditions are commonly found in the lee of bergs or hummocks in bays or fiords, at the mouths of glaciers, and among islands. These breeding requirements are evidently so important that they may apparently take precedence over food requirements, for seals are often numerous in localities where the feeding is poor, yet scarce or even non-existent where conditions are good—as in Hudson Bay.

By the end of October a hundred or more ringed seals can be observed at one time lined up on the ice beside their holes, some only a yard apart, others at half-mile intervals. At the outset a seal weakens its selected pieces of ice by prodding them with its snout or bashing them with its head, and then applies itself to gnawing industriously at them with teeth that are set far enough forward in its jaw for it to gnaw almost as efficiently as a rat. As winter draws on and the ice thickens, so the seal's persistent gnawing from below raises the edges of its air-hole into a miniature dome of ice 3 inches or 5 inches above the water level. With a diameter of from 1½ to 2½ inches it is large enough to contain the seal's muzzle. However, in low temperatures, ice forms so rapidly about the air-hole that according to some reports a seal's snout may become frozen to the ice, with the result that it either suffocates or starves to death. Eventually the dome is buried under several feet of snow, but no matter how thick the snow covering, sufficient air percolates through for the seal to breathe. The hole below the vent is continually enlarged by the seal's coming and going, and as the depth of the ice increases to 2 or 7 or exceptionally 10 feet, it acquires the shape of a conical inverted bowl or bell, widest at the bottom.

Once the ice is permanent, every seal is a prisoner within the area it can fish between visits to one or other of its eight or ten breathing holes or *aglos* as the Eskimos term them; though prisoner is perhaps hardly the right term, since a seal can swim underwater at a rate of some 600 yards a minute, dive to depths of 300 feet, and remain submerged for certainly 7 or 8 minutes and perhaps as long as 20 minutes. However, the more aglos it can keep open the greater the potential area of its fishing waters; while if hard-pressed for air it can make use of another seal's aglo. Moreover, twice a month, with the change in the moon, high tides produce cracks in the ice a few inches or feet wide. Year after year these tidal cracks occur in the same localities, and many seals desert their regular aglos to open up new holes in the thin ice forming over the cracks.

How does a seal locate its aglos? In considering the everyday problems confronting Weddell seals in the Antarctic it has not been possible to determine what technique they employ to locate their prey or, for that matter, their breathing holes during the permanent winter darkness under the ice. In the Arctic spring Stefansson observed that, when the overlying snow was removed from the dome of a ringed seal's aglo, the occupant was frightened away from it by the daylight filtering through the vent. He therefore suggested that to a seal swimming beneath moderately level ice the aglos might appear slightly illuminated in comparison with the surrounding ice. But what physical features are there beneath the ice by which a seal can remember the position of its aglos, and how can it find them when a blizzard is raging over the ice or during those prolonged moonless and overcast periods characteristic of the polar winter, when it is as dark at midday as it is at midnight? As Freuchen noted in *Arctic Adventure*:

> When the snow lies meter-high over the ice and it is the dark time, it cannot be the light that guides it: I have often examined the underside of the ice, but there was no noticeable mark. Yet it seems that without hesitation seals swim from one hole to another. Nor has the spotted (common) seal any difficulty in finding the hole of the fiord (ringed) seal.

Were it not that seals compress their nostrils when submerged one might hazard a guess that they smell out their aglos, for ringed seals have a musky body odor; and towards the end of the winter old bulls, in particular, permeate the aglos and the air for some distance around with a pungent "chemical" stench.

Although ringed seals obtain the bulk of their food in the upper 120 feet of water, and certainly not at greater depths than 300 feet, they too have had to solve the problem of detecting it in winter conditions of near or total darkness under the snow-covered ice. At this season they probably feed mainly on large planktonic crustacea, many of which exhibit a brilliant luminescence, but such fish as cod and sculpins are also taken, and these are perhaps detected sonically. There is also the problem of sleep. If they sleep in their aglos they must do so very briefly or become frozen to them, except perhaps where these are overlaid by deep snow and sheltered from the wind by a stranded berg or heavy pressure-ice at the head of a fiord. On March 22, for example, Frederick Jackson (who lived three successive winters on Franz Josef) found a large ringed seal, which he believed had been frozen out of its hole and worried by foxes, lying dead on the ice. Yet during the spring and summer, when they can enlarge their aglos sufficiently, or take advantage of leads or tide-cracks, to haul out, ringed seals pass much of their time basking or sleeping on the ice for 15 or 20 or even 40 hours at a stretch; and they probably fast for lengthy periods throughout the breeding and molting season from April until June or July, when nine out of every ten killed have empty stomachs. Obviously they cannot sleep on top of level fields of exposed ice during the winter, because they would not survive the low temperatures, and in any case their aglos would probably freeze over and prevent their return to the water. Even in the spring a seal, taking a nap on the ice in frosty weather, may sleep so long that the crack through which it surfaced closes together or freezes over. In these conditions, however, Freuchen observed that a seal might succeed in digging down through snow and ice with surprising rapidity, shoveling the snow behind it with its flippers in the manner of a dog digging. One's guess is that ringed seals, like Weddell seals, sleep very little in the winter. Eskimos apparently believe this to be the case, for they employ the identical word *sinik* for both sleep and spleen, in recognition of the fact that during the summer when the seals are sleeping a great deal their spleens are very large, whereas during the winter they are small. I have not been able to discover what relationship there can be between the size of spleen and degree of sleep, but it is also a fact that the spleens of Weddell seals are enormously enlarged when they begin to lie out on the ice in the spring, preparatory to pupping.

It is late in February or in March—later in the High Arctic—that the older ringed seal cows prepare birth-chambers, though the majority of the pups will not be born until early April, and some not until mid-May. Enlarging the vent of one of her aglos with her strong claws to a diameter of 2 feet, the cow squeezes herself out on to the ice and sets about excavating a chamber in the deep snow-cover. The chamber is just large enough to admit her body, though she can barely raise her head, for it is only 4 or 5 feet long, 3½ or 4 feet wide, and from 2 to 2½ feet high. The pup, 18 inches long at birth, lies on the ice at the extreme end of that part of the chamber farthest from the opening into the aglo, for the very good reason that it cannot swim for the first two weeks. If it falls into the water, only a couple of inches below the ice, its long downy fur quickly becomes saturated, and the cow, who is never far away, has great difficulty in heaving it up on to its platform before it drowns. Since by this time the ice through which the aglo is "drilled" may be 10 feet thick one wonders how the cow manages to heave herself up through the narrow funnel of the aglo into the chamber in order to suckle the pup; but she does, and the heat of her body, together with the relative warmth of the water in the aglo, not only raises the temperature of the chamber for the pup but, by melting a veneer of the snow which subsequently freezes into a strong crust, also strengthens the structure.

By the time the pups are a month old, the spring thaw has begun in the Arctic. Streams are flowing down the fiords and there are pools on the ice. The sun's heat is now strong enough to melt the snow roofs of the pups' chambers, and occupants of those exposed to its direct rays must leave before this catastrophe reveals them to bears, wolves and foxes—all ravenous at this season. By June as many as eight cows, together with their pups, may be seen lying around a single hole or beside a tide-crack. The pups amuse themselves, and also learn the fundamentals of seal life, by diving time and time again into the water. The cows are perpetually restless, scratching with their hind flippers at the lice that cluster in hundreds in the folds of their skin, rolling over to cool sides that have become too heated in the sun, lying on their backs and fanning themselves with both front and hind flippers. At every change of position they raise their heads to scan the ice or test the wind for bears. In July, they commonly haul out in the early mornings or forenoon, after feeding on shrimps or crustaceans, and groups of twenty or a hundred in favored localities, lie beside the tide-cracks which extend for miles, until midnight when they return to the water. The cows continue suckling the pups for as long as two months, because the majority are weaned on the stable land-fast ice which will not break up early in the summer, in contrast to the unstable pack-ice on which the harp and hooded seal pups are born. Indeed there is some evidence that the older ringed seal cows are experienced enough to select ice which will not break up until late in the season, and that the pups of these cows are suckled longer and grow larger than those born on less stable ice. This, if correct, is a significant confirmation of the influence of the ice factor in the ringed seal's life history.

13: The Walrus Herds

In addition to the true seals there are about 125,000 walruses in the Arctic, three-quarters of which belong to the Pacific race inhabiting the Chukchi and Bering Seas, and the bulk of the remainder to the Atlantic race (distinguished primarily by smaller tusks) ranging from Labrador and Hudson Bay to Baffin Bay and the more easterly of the Canadian Arctic islands. A third geographical group is located off north-east Greenland; a fourth in the Barents and Kara Seas as far north as Svalbard and Franz Josef Land; and a fifth in the Laptev Sea farther east along the coast of Siberia.

The existence of the herds depends upon a guaranteed, all-the-year-round supply of shellfish; and their daily movements and seasonal migrations are governed by the extent to which the unevenly distributed shellfish banks are frozen over or are affected by the variable drift of the pack-ice. The banks must be extensive and highly productive because walruses not only associate in large aggregations comprising hundreds or thousands of individuals, but also enjoy collossal appetites; they must not lie deeper than 250 or 300 feet, which is the maximum depth to which a walrus can dive; and they must be situated, whenever possible, near floe-ice on to which the herds can haul out when not feeding. Walruses are powerful swimmers, cruising just beneath the surface at a speed of about 6 knots by striking alternately with front flippers in a kind of reverse breast-stroke, while the hind flippers sweep from side to side with a rudder-like motion. Nevertheless, they apparently find it necessary, both at their feeding banks and when on migration, to rest and sleep for lengthy periods, unlike other seals which sleep in short snatches at any time. They rest on the floes when the sea is rough, and they haul out on to them on calm sunny days at any season of the year, to bask or sleep in pods of twenty or thirty to a floe for hours at a time. Sprawled on their backs or piled one on top of another, they may lie in one place for so long that the snow melts beneath them. Indeed the bulls, whose layer of blubber (thickest during the winter) apparently renders them impervious to the most extreme cold, will sleep on the ice for 36 hours at a stretch in upwards of 70 degrees of frost with a 20-mile per hour wind blowing.

This rest factor probably explains why the largest known herds have always frequented the Bering and Chukchi Seas. There, off the north and south coasts of Chukotsk peninsula, vast acreages of shellfish banks lie in shallow waters which extend as far as 15 or 20 miles offshore; and there the pack-ice is present until exceptionally late in the summer, so that with winds and currents ever drifting the floes hither and thither, resting platforms for the feeding herds are available over a wide area. So long as the ice remains in the vicinity of the banks, the walruses crowd together on the floes throughout the summer. If the pack-ice disperses they move to other banks where ice is present. Likewise, if contrary winds blow the floes away from the polar coast of Alaska the herds summering in those waters rarely come inshore; while in those autumns when the advance of the ice into the Bering Sea is retarded, they are likely to remain near the edge of the pack rather than continue their migration to their traditional wintering waters further south. Thus by local or long-distance migrations they remain in contact with the ice throughout the year—with certain exceptions to which we shall refer. They are not, however, passively dependent on the drift-ice for public transport, as was formerly believed, for they only ride on the floes if these are drifting in the right direction. If in the spring, as early as March perhaps, the northerly recession of the pack-ice and its coincidental break-up have been retarded by low temperatures, the herds migrating north to the summer banks continue to press on if the pack is negotiable, swimming alternatively on the surface or just beneath it, curving in and out of small floes, diving to pass beneath large ones, and hauling out on warm days to rest on the ice. Similarly, in exceptionally warm autumns, when the pack-ice is late in invading the Chukchi Sea, those walruses that have summered at the banks off Wrangel Island do not await the arrival of the drifting floes on which to hitch a ride, but swim across the open sea to Bering Strait. But in so doing they exhaust themselves and haul out on to the first suitable beach, notwithstanding that this may be near a native settlement. Because of their need to surface for air at 10-minute intervals the walruses must migrate from their summer feeding banks when the autumnal layer of new ice eventually becomes too thick for them to keep open permanent breathing holes, and retreat to areas where there are more or less permanent stretches of open water in which they can dive freely to the banks. Thus, the Pacific herds, summering in the east Siberian sea and off the northern coast of Alaska, must embark on a long-distance migration, in order to pass through Bering Strait before the ice closes it by the middle or end of October, and penetrate those parts of the Bering Sea in which the pack-ice reaches its most southerly limit. Less than 40 miles wide, Bering Strait serves as a funnel for perhaps the most spectacular migration of marine mammals and birds to be seen anywhere in the world, since it includes not only the thousands of walruses, but also those of four species of seals and five species of whales, together with millions of auklets, harlequins, scoters and eider duck. In July 1881 the U.S. *Corwin* cruised for hours along the edge of the Alaskan ice past an almost unbroken line of "tens of thousands" of walruses hauled out on the floes; and early this century, when a single herd of walruses might take half a day to pass an Eskimo settlement during its migration through the Strait, the American Frank Dufresne (quoted by E. W. Nelson) could describe how, when the sun burnt through the banks of fog rolling over the ice-fields off the north coast of Alaska:

> The vast ice-plains ahead are seen to be black with huge grotesque creatures. Walrus—5,000 of them stretched sleeping on the white ice. In pods of from 40 to 50 they lie heaped against each other almost as far as the eye can see. Then, with celerity unbelievable in such gross creatures, they tumble off the ice; the sea is filled with bobbing, tusked heads. They roar hoarsely, or give vent to the peculiar whistling neigh suggestive of the name "sea-horse" bestowed on this odd animal by the navigators of an earlier day.

Today, the walrus hunters of the Bering Sea islands must often voyage long distances in their power-driven *umiaks*—formerly the women's boats—to reach the decimated herds far out on the ice-fields. Yet, in the eighteenth and nineteenth centuries, when sealers and whalers were engaged in slaughtering several million of them, walruses were reported to have crowded so densely in "hundreds of thousands" on islands in the Bering Sea that the constant traffic of their heavy bodies denuded some of the smaller islands of all their vegetation.

In other sectors of the walruses' range, such as north-west Greenland, where the inshore ice may already be thick enough by October to prevent them diving to their banks, a relatively local coastal migration may suffice to take the herds to the outlet of some island-studded fiord, where strong currents racing between the skerries prevent the formation of thick permanent ice. Or they may have to retreat farther and farther from the coast, as the sheet of land-fast ice extends seawards, until by the New Year they may be 10 or exceptionally 100 miles off the land. Out there a broad zone of sea-ice is in constant motion and is continually being broken up by the waves and the swell into small floes, with extensive leads of open water. However, these relatively deep-water banks must place the walruses at the disadvantage of having to expend a maximum of time and energy in diving to their limit in order to collect their mollusks; and whenever high tides crack the inshore ice and break off floes, the walruses move in again, often ranging long distances under the thick ice.

A walrus' teeth and tusks are not adapted to keeping open breathing holes in thick ice; but he can do this in ice that is not more than a few inches thick, though according to Freuchen he will not attempt to break through snow-covered ice which appears black from below. More than 300 years ago that admirable Dutch seaman, Van de Brugge, observed that the walruses off Novaya Zemlya would position themselves close beneath the ice and, arching their backs, suddenly straighten them and strike the ice with the crowns of their heads. A walrus will in fact repeat this action several times until he has shattered the solid ice or the new film that may reform over a polynia while he is submerged; and, as we have already noted, the heads of some walruses bear scar-tissue. Such is their impact on the ice that the lines of fracture radiating from their breathing holes are much more marked than those from any seal's hole, and may extend over an area of 60 or 70 yards from the cupola. A number of walruses may combine in opening up a breathing hole, and in the middle of October, Elisha Kane, the mid-nineteenth-century American explorer, watched five "rising" at intervals through the ice in a body, and breaking it up with an explosive puff that might have been heard for miles. If the ice is too thick to be shattered by his head, a walrus is reported to employ an alternative technique. Tapping at the ice with his tusks while floating on his back, he works round and round and gradually chisels out a hole, though in view of the fact that his tusks are slightly recurved, and may also curve inwards or outwards at their tips, this would appear to be an awkward operation for him.

Taking into account a bull walrus' immense body weight of a ton or even a ton and a half, the strength of his tusks, from 1 to 3 feet in length, is prodigious; for though those of a Pacific bull may be 10 inches in circumference at their base, they average no more than 3 or 4 pounds apiece in weight, and rarely exceed 12 pounds in the largest bulls. With the aid of tusks and flippers a walrus hauls himself out of the sea by crawling up on to the submerged apron of an ice-floe, and hooking his tusks over the edge of the exposed ice and contracting his neck muscles, pulls himself up the steep edge high enough to place one flipper on top of the floe. Then, taking the strain with the tusks, he heaves forward on his flippers, whose rough palms, encased in horny skin a quarter of an inch thick, prevent him from slipping on the smoothest ice. Almost a hundred years ago Edward Belcher, watching a bull hauling himself out of the water and projecting a third of his mountainous carcass on to a floe, described in *Last of the Arctic Voyages* how:

> It then dug its tusks with such terriffic force into the ice that I feared for its brain, and, leech-like, hauled itself forward by the enormous muscular power of the neck, repeating the operation until it was secure. The force with which the tusks were struck into the ice appeared not only sufficient to break them, but the concussion was so heavy that I was surprised that any brain could bear it.

Nevertheless, although a bull also uses his tusks to hook himself over hummocks on the ice-fields, he is careful not to knock them against rocks when hauling out on a beach, employing his fore-flippers solely for this purpose, and waiting patiently for the surfing rollers to boost him up the beach in stages.

Where strong currents prevent thick ice forming even at mid-winter, as is the case along some stretches of coast off both west and east Greenland, a family or small pod of walruses may be able to maintain a permanent winter home in a chain of breathing holes up to 12 feet in diameter; for no matter how thick the ice, some walruses are extremely reluctant to leave productive shellfish banks, just as others return to them as soon as there is any break-up in the ice. Their breathing holes are always much more numerous in the spring than in the autumn, and early in April one year Christian Vibe, the Danish zoologist, observed that, after a very severe frost, when the ice had extended seawards from coastal banks off west Greenland, the wintering walruses were faced with the alternatives of either leaving the area or of keeping the ice open as long as possible. In these circumstances they chose the latter alternative, with the result that glass-like cupolas, 3 feet in breadth, were formed around their holes by the condensation of vapor on the ice whenever one came up to blow. Bearded seals produce the same effect around their small breathing holes, and these cupolas, when reflecting sunlight or moonlight, are conspicuous at a great distance.

Despite their millennia of inherited experience, walruses are sometimes at fault in their timing of ice conditions; and an individual may become imprisoned in a bay or fiord because it has stayed too long in a small polynia and allowed the ice to form around it so quickly that it cannot break through to open water. If the numbers of shellfish on the sea bottom at this point are limited, the walrus is doomed to die of starvation, unless he can heave himself out on to the ice and set off in search of another polynia. Walruses are reputed to be able to smell water at a great distance, and there is some evidence that one trapped in this manner will haul his immense bulk over the most formidable ice-hummocks blocking his direct route to the nearest water rather than detour around them; but one doubts whether many survive to reach open water in view of the great weight their bone structure has to support when on land.

During the winter some walruses sleep beneath their blowholes, rising intermittently to hook their tusks over the edge of the ice and take a few breaths before sinking to sleep again. But most walruses sleep in leads and cracks, either floating on the surface on backs or bellies, lifting their heads every few minutes to breathe, or supported by the unique pair of air-sacs placed beneath the throats of the majority of adult walruses. The upper part of a walrus' oesophagus is very muscular, and when these sacs are inflated from the lungs the air is retained in them by constrictors. With their support a walrus can float in an upright position, with part of the sacs and the tip of his muzzle above the surface; and bulls in particular may sleep soundly in this posture, and long enough for a thin layer of ice to form around them. Although circumpolar in distribution, walruses do not normally range far beyond the southern edge of the pack-ice; and the fact that some have been able to accomplish long oceanic voyages to such foreign shores as those of Iceland or the British Isles, at least 600 miles from the nearest ice, may perhaps be attributed to their being equipped with these air-sacs, which enable them to rest, and also sleep, without hauling out on to ice or rocks, providing that the sea is not too rough. One wonders, though, how a solitary calf, sighted at the Faroe Islands on Christmas Day 1934, had been able to complete a voyage of not less than 500 miles without food when at most only nine months old.

The walruses' breeding requirements do not, curiously enough, influence their seasonal distribution in any way because they do not haul out on to land and assemble in immense rookeries in order that the bulls may collect around them harems of cows, as elephant seals, sea lions and fur seals do; but mate and also calve on the pack-ice, usually while actually on migration far out to sea, during the months of April, May, and especially June. There is, however, some circumstantial evidence that cows of the Atlantic race may formerly have hauled out ashore to calve on islands to the north and south of Nova Scotia. These were often ice-free in the spring, whereas in higher latitudes heavy pack-ice must always have prevented walruses from hauling out until July or August, after the breeding season. It is true that walruses do establish large rookeries, or *uglit*, on beaches; but these are predominantly aggregations of bulls; for a feature of walrus society is that a large percentage of the bulls are separated from the herds of cows and immature beasts throughout the year, though especially during the summer months when some do not even migrate to the same feeding grounds as the cows, or may actually not migrate from their wintering area. The *ugli*, or rookery, on Round Island—one of the Walrus Islands in Bristol Bay on the west coast of Alaska—not only serves, for example,

Walrus

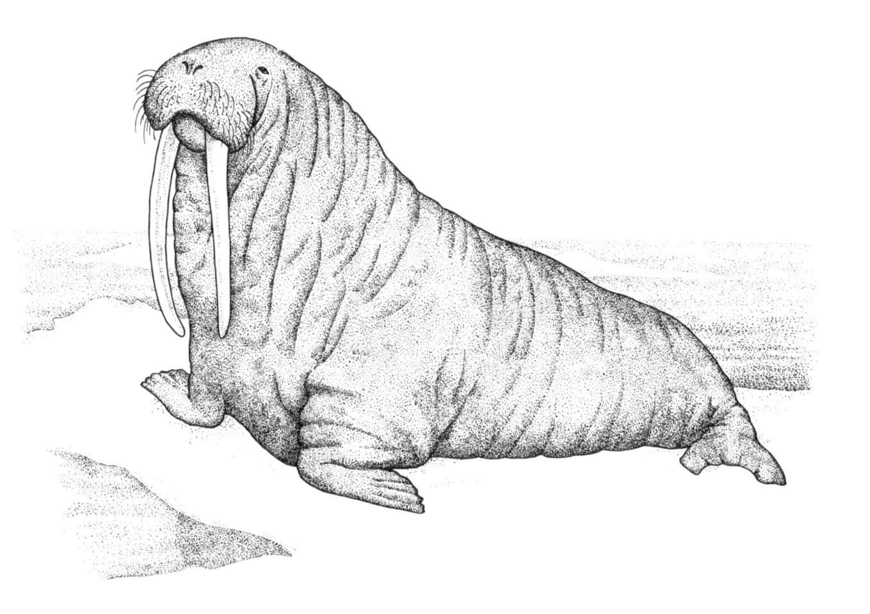

as a summer retreat for that proportion of the bulls not migrating north in the spring, but is actually situated to the south of the cows' wintering area. This sector of the Bering Sea has indeed been inhabited exclusively by bulls during the summer months for as long as records are available. Although this ugli includes numbers of the younger bulls, and no doubt some aged bulls no longer able to command cows, among its 2,000 or 2,500 "bachelors," the astonishing fact is that large numbers of them are prime bulls. Segregation of the sexes for prolonged periods is a common feature of mammalian herd society; but in view of the strong herd ties manifested by all ages and sexes of walruses—barren cows and young bulls protecting orphaned calves whose mothers have been shot: older bulls protecting younger bulls and supporting wounded cows on their backs —it seems remarkable that this segregation, which is also observed by herds of barren cows, should be so marked in their society. Traditional bull uglit have been occupied in the knowledge of Eskimos for centuries, and Round Island has been familiar to Europeans for more than seventy years. What can be the reason for this phenomenon? Is it associated with the walruses' mating habits? In any population of walruses there are apparently equal numbers of bulls and cows; but since one bull may mate with as many as five cows, large numbers of bulls will be unable to mate every year. Rather than adopt the harem system of other large seals, walruses have apparently found it preferable to limit the number of potential mating bulls by this voluntary omission on the part of several thousands of them to accompany the cows to the summer banks.

Although a herd of walruses will alternate freely between a number of uglit in one locality, in order to move to fresh feeding banks or to take advantage of the shelter one or other affords from heavy surf or strong onshore winds, it returns year after year to the same ones providing that it is not discouraged by persistent persecution. All uglit possess certain physical features that do not occur universally on Arctic coasts. Shallow beaches are not suitable, because they involve the crossing of extensive flats at low tide; nor are beaches of shingle or loose sand, because these do not afford a firm hold for a walrus' flippers. The ideal beach shelves gradually down into a fair depth of water, enabling the walruses to haul out—or return quickly to the sea if necessary—at both high and low tide. This conformation also enables them to take advantage of the rollers when hauling out of a rough sea, and to rest at the edge of the sea for lengthy periods on calm days, before heaving themselves up the beach. Sloping rocky shores are preferred, rising in broad terraces which can accommodate large numbers of walruses; and these should be striated with ledges and crevices on which the walruses can obtain a purchase with their flippers, even if the rock has been worn smooth by glaciation and water erosion. In Hudson Bay and Strait, where much of the shoreline is composed of unsuitable flat limestone rock, most of the uglit are sited on small islands or on the crystalline rock of prominent headlands; while those on the narrow rock or gravel beaches in the Bering and Chukchi Seas are also based near high island promontories or projecting coastal headlands, which could serve as landmarks to the homing walruses. In many instances the uglit are backed by steep walls of cliff, as is dramatically the case on Round Island where a 300-yard strip of cobbled beach, only 30 feet in breadth, lies hard up against the base of towering granite crags. Possibly this type of beach is selected as affording protection against the attacks of polar bears; possibly because it conforms to the walruses' obsession that whether they haul out on ice or shore they must invariably pack together as closely as possible. It is significant that Kotzbue Sound, which lacks both summer ice and the alternative, a narrow beach with steep cliffs as backdrop, has apparently always been shunned by the herds when migrating through Bering Strait.

The bulls may even haul themselves up steep cliffs to a height of 60 or 100 feet above the sea, and venture a considerable distance inland. That such a mountainous beast can scale rocks is due to the peculiar structure of his flippers, of which the front pair are more than 2 feet in length and 15 inches broad, and the fan-shaped hind pair, though 6 inches shorter, immensely broad, measuring from 2½ to 3 feet when fully extended. Basically, they are flares of skin stretched over bone and muscle, with free movement from the elbows; and the hind pair can be rotated forwards and are mobile enough for a walrus to be able to scratch the base of his neck with his claws. This jointure enables him to raise the hinder part of his body and support it on the palms of his flippers, permitting him greater freedom of movement on land than the smaller seals—though he must pause to rest every few yards—as he advances first one front flipper and then the other to it, follows up with his hind flippers, and finally humps forward his massive mid-body. Despite his immense bulk a bull walrus is neither ponderous in his movements nor slow in his reactions when occasion demands, heaving, shuffling and humping along at surprising speed while, when fighting, his thrusts at a rival's neck and shoulders are also surprisingly swift as he lunges forward with the upper part of his body, head craned to one side, and strikes with sufficient force to draw blood. Normally, however, the blunted tusks do no more than bruise an adversary's exceedingly tough hide, rarely penetrating to a vital spot through the 3 or 4 inches of blubber. Although old bulls dispute over desirable resting places, and young bulls strike at one another playfully, there is no evidence that fighting results in any significant mortality among walruses. As in most mammalian societies, excessive fighting, with its genocidal threat, is lessened by the employment of warning postures. Rearing up on his front flippers to a height of 5 feet, a bull walrus expands his great chest and, raising his head almost vertically so that his tusks project horizontally, blows threateningly through his lips; and perhaps follows up this warning with a number of tentative jabs with his tusks.

Although some walruses molt while on the pack-ice the majority of both sexes do so at uglit, where this operation can probably be conducted more satisfactorily than among the ice-floes, when they are constantly in and out of the water. The individual walrus completes its molt in a couple of weeks; but the overall molting season extends from the middle of June, when the bachelor bulls begin, to the middle of September. These are the months when the coastal ice breaks up, and therefore those when most herds haul out ashore, particularly in those areas where there is no pack-ice; and a herd of bulls may remain permanently ashore for as long as six weeks at a sheltered ugli. The molt is a period of almost continuous sleep and fasting for all except the nursing cows and young beasts, though this prolonged inactivity does not prevent them achieving their maximum growth during the summer and autumn. They sleep soundly, sighing deeply and snoring, on their sides or backs with flippers in the air, one using another as a pillow, and if one raises its body laboriously this act is sufficient to evoke a series of complaining groans and grunts from its neighbors. With relatively small heads almost hidden by the heavy folds of rust-brown skin rolling over their necks they much resemble jumbles of boulders. Wheatears and snow buntings hop about among them and on their backs, picking up parasites.

Even on a warm day, when a herd of bulls may be waving their flippers in order to cool their overheated bodies, they still, however great their discomfort, remain closely huddled together. So too, a herd of bulls at sea invariably crowds on to the smallest floe that will accommodate all its members and, though there may be space to spare, packs one part of the floe as densely as when constricted by the narrow confines of a strip of beach. Exceptionally, as many as a thousand bulls may haul out on to a single floe, and considerable jostling and stabbing with tusks ensues as the floe gradually sinks 4 or 5 inches into the water beneath its 1,000 tons burden. Some are pushed off into the sea; others lie awash rather than leave the floe which, however, eventually capsizes and tips the lot of them into the water.

Why do walruses associate in such large herds, and lie out in such congested masses? They do not, as we have seen, congregate in order that the bulls may collect large harems of cows; nor are they menaced by any predators whose attacks would result in losses serious enough to render herding advantageous. It has been suggested that herding assists the younger members to learn the traditional migratory routes and the location of the various shellfish banks, or that it serves to keep individuals and family pods in contact during those long foggy spells characteristic of many parts of the Arctic, and also during the four or five months darkness of the polar winter. This latter purpose is, however, well served by the walrus' voice, whose strong bellowing and barking can be heard at a distance of several miles on calm days, and is the loudest voice in the Arctic. In fact neither of these advantages would be basically important enough to have initiated such behavior. More probably herding was the natural consequence of the seasonal concentration of immense numbers of family pods at productive shellfish banks in the vicinity of suitable uglit; while their massing together is perhaps explained by the awkward conformation of their tusks, for if a walrus wishes to sleep—which it does very often—it can only do so by lying on its side or back, or by resting its tusks on the prostrate body of a companion; though it must be admitted that this is not a very convincing explanation.

The calves are, as we have seen, born on the pack-ice and most frequently while the cows are on migration to the summer banks; but although they are initially awkward at swimming and dislike submerging, their birth does not delay the migration of the herd, for they ride on their mothers' backs or cling to their breasts. For the first few months after birth they grow very slowly, reaching a length of only 5 feet and a weight of 450 pounds by the end of their first year, and less than 7 feet and 750 pounds by the end of the second. Although the calves may stray temporarily from their mothers when they are a year or so old the ties between the two are virtually indissoluble until the calves are eighteen months or older, and the cows continue to suckle them assiduously throughout this period. This prolonged lactation, vastly longer than that of any other seal, is probably necessary because a walrus can only obtain its specialized shellfish food with the aid of its tusks; and although the calves' canine teeth begin to erupt between the age of two and five months, it is two years before they protrude as tusks 3 or 4 inches long. The constant surveillance of the calves by the cows during this prolonged weaning must also be of considerable survival value, since the walruses' rate of reproduction is very slow and more than a third of the population is immature.

Creatures of such size and strength may not be significantly threatened by any natural enemies, and the confidence with which they settle down to sleep on the ice or on beaches for hours or days at a time confirms that this must have been the case for a very long time. On the other hand, since they are manifestly intelligent animals, their, in general, total lack of caution when hauled out is difficult to understand when one recalls their centuries of persecution by sealers and whalers. There has always been the natural wastage in annual losses among calves crushed during storms and pressure in the pack-ice, or by bulls stampeding over the beach to the sea when panicking at the scent of men they cannot see or when buzzed by lowing-flying aircraft. One presumes, too, that killer whales have always taken their toll. According to Bering Strait Eskimos a killer will attack a walrus in the same way as it does a narwhal or beluga, hurling it right out of the water; or alternatively, seizing it by the upper lip, dragging it under water and drowning it. The mere presence of a pack of killers in inshore waters is said to be sufficient to cause a herd of walruses lying out on the rocks to stampede. About the year 1936 a large herd was reported to have been driven ashore in such panic by killers that, while hauling out on to the beach through the surf, they piled up one on top of another, with the result that more than 200 were crushed or smothered.

Yet the fact is that there does not appear to be a single *definite* record of a walrus being killed by one of these whales, nor even attacked by them in their Atlantic and Canadian Arctic sectors where, in Peter Freuchen's experience, walruses could be seen lying undisturbed on their floes or swimming unconcerned among them when seals and narwhals were terror-stricken by the presence of a pack of killers. Some Eskimos indeed believe that killers actually fear walruses, and in order to scare away a pack of killers the Etah Eskimos of north-west Greenland would grunt and bellow, and dip their harpoon heads into the water in simulation of walrus tusks. There is at least one authentic record of a bull walrus who, when a pack of killers approached his floe, plunged into the water among them and subsequently surfaced with his tusks implanted deep in one's back. And cows, whose calves have been swept off their backs by killers, are also reported to have inflicted mortal wounds on their aggressors.

More frequent and successful attacks on walruses are made by polar bears, whose geographical distribution ties up with those of the various populations of walruses, as it does with those of ringed seals. A recent survey by Russian zoologists suggests that bears may prey more commonly on walruses than has been supposed, for almost a quarter of the stomach contents of bears shot off Franz Josef Land, mainly during the period March to May when the herds are migrating, comprised the remains of walrus, including those of adult bulls. Nevertheless, if bears do in fact present any significant threat to walruses, it must be predominantly to the cows and calves, and perhaps the younger bulls when they are in their soporific molting condition. In the water a walrus normally has nothing to fear from a bear which, though a strong swimmer for a land animal, cannot match the maneuverability of either walrus or seal. He is indeed so relatively helpless that half-a-dozen young ringed seals—his favorite prey on the ice—will gambol around him, nipping at his flanks and causing him to hasten to the nearest floe. However some young walruses, and also seals, are captured by bears swimming on their backs underwater and seizing them from below when they pop their heads up to breathe or look around. Despite the Russian reports, however, one cannot believe that bears find walrus-hunting very profitable. Alwin Pedersen, the Danish trapper and naturalist, indeed asserts that they are so afraid of walruses that they will not venture into the sea if there is a herd in the vicinity; while according to Frank Illingworth, a dozen or more walruses will gang up against a marauding bear, as they will also do against a killer whale. One of Illingworth's Eskimo companions saw three cows drive a she-bear into the sea and gouge her to death with their tusks after she had killed a calf, and he describes in *Wild Life Beyond the North* how he himself watched a bear stalking with the utmost care the smallest of three bulls lying out on rocks at the edge of the sea. When the bear made his final lunge, the largest bull turned on him:

> Exactly what happened next I cannot say. But in a matter of seconds the bear lay bleeding from a mortal wound and the old bull, roaring angrily, lifted itself on its fore-flippers and surveyed the scene of its victory before taking to the water. When I examined the bear I found a jagged wound in the neck from which its life-blood flowed over the rocks.

But even the largest he-bear, standing 10 feet or more when erect and weighing three-quarters of a ton, must be very hungry to tackle a mature bull walrus 10 or 14 feet in length and no less in girth, despite the immense strength a bear displays in destroying explorers' caches constructed of the heaviest beams and rocks. A bear and bull walruses on the same floe normally ignore each other, though the latter are of course greatly inferior in agility when out of the water.

There cannot be said to be any significant ecological competition between walruses and polar bears, though some bull walruses, which have adopted carnivorous habits, feed on such carrion as the carcasses of whales, and these form vital reserves of food during the hungry months of the long polar winter, not only for bears but for many other Arctic predators from ivory gulls and ravens to foxes and wolves. Some of these carnivorous bulls also kill ringed seals, probably young bearded seals, and such powerful beasts as narwhals and belugas, and capture large numbers of tom-cod when they are swimming in shoals at the surface. Both Pedersen and Freuchen state that seals and whales are afraid of walruses, avoiding localities where there are uglit and the bays they frequent during the summer months. Though walruses have been reported breaking through the ice in attempts to get at basking seals, they normally hunt them in the water; but since the seals are both faster and much the more agile, the walrus must capture them by stealth, employing a bear's tactics. Swimming on his back, he glides beneath the seal at the moment when the latter is lifting its head out of the water to breathe, and seizing it behind the flippers, strikes with his tusks, ripping open the seal's breast. Then, holding the seal between his flippers (as a cow does her calf), the bull makes for the nearest floe. There he opens up the seal further with his tusks and gulps down pieces of skin and particularly blubber. However, the limited space between his tusks, and their placement at the corners of his jaws, restricts the size of the pieces, with the result that the greater part of the carcass is usually left untouched.

Why are a minority of walruses carnivorous? Difficulty in obtaining an adequate supply of shellfish could be one reason, and would particularly affect aged bulls, some of which evidently shun the society of the herd and are to be found alone or in small pods on the southern periphery of the herd's range. Once such a bull—unable, because of some physical disability perhaps, to procure shellfish in sufficient quantities to satisfy his needs—has killed and eaten a seal or maybe first sampled carrion, he may thereafter continue to prefer carrion or seal meat to mollusks, and deliberately prey on seals. There is some evidence that hunger can change a walrus' feeding habits. Seal-hunting walruses are, for example, reported to be especially common off north-east Baffin Land, where shellfish banks are not only thinly distributed but also often inaccessible because of heavy ice, and it is possible that this factor accounts for walruses becoming seal killers in some parts of the High Arctic. Herds on migration also often experience difficulty in obtaining shellfish. In the days when the north-west Greenland herds migrated south past Pond's Inlet on the Baffin coast it is reported that they were so ravenous that they would "leap" up on to the ice and tear at the carcasses of narwhals (which they had smelled at a great distance) when the Eskimos were actually engaged in flensing these. On the other hand the Eskimos of Alaska and also the Chukchi believe carnivorous walruses to be "rogues" which were orphaned before being weaned and driven from the herd. Though such a theory does not accord with the remarkable herd solidarity of walruses, nor explain why only bulls are carnivorous, the contention is that since these rogues were too young to excavate shellfish when they were orphaned, they were obliged to eat anything they could pick up in the way of fish, "corpses of drowned men, and carcasses of animals." Then, when they were older and stronger, they would begin to attack seals and subsequently the cows and calves of their own kind, tearing them apart with their tusks; and according to the Chukchi, when a rogue hauls out at an ugli the resident walruses, repelled by the unnatural odor of a flesh-eater, immediately make for the sea.

But 99·9 per cent of walruses are shellfish eaters, feeding mainly on such bivalves as mussels, cockles and clams, but also on whelks and other sea-snails, together with sandworms, sea-cucumbers, starfish, shrimps and hermit-crabs. The bulk of their food, however, comprises two species of clams, both of which are normally deeply buried among stones or in holes. One, the small though thick *Saxicava arctica*, with finger-like foot and long syphon or breathing-tube, affixes itself by its bissal cord to the root-anchors of large seaweeds. The other, *Mya truncata*, with tongue-like foot and soft chalky shell as large as the palm of a man's hand, buries itself from 3 to 12 inches deep in gravelly mudflats, where it is also sought by birds and foxes. Both live predominantly at depths of between 130 and 265 feet, though *Saxicava* is also found in only 50 feet of water. Pacific walruses, who appear to feed mainly in shallower seas, prey extensively on a third clam, *Clinocardium nuttali*, and also take large quantities of *Neptunea* and *Sihpo* snails at depths of from 20 to 100 feet. Walruses are selective in their choice of particular mollusks. When high tides break up the winter ice off Thule, enabling the Greenland herds to move closer inshore and feed in depths not exceeding 125 feet, a cockle, *Cardium groenlandica,* is preferred to clams or mussels; while, in certain localities further south, Christian Vibe found that the herds were feeding almost exclusively on the large fat feet of this cockle and those of another, *Cardium ciliata,* until the winter extension of the coastal ice seawards forced them to feed in 250 feet of water at too great a depth for cockles, but suited to clams.

The techniques employed by walruses in obtaining their shellfish are various and debatable. The ideal banks are naturally situated in shallow waters in which the walruses can collect their shellfish with the least expenditure of energy and time, and so, when the landward ice begins to break up in summer, they move into coastal waters to feed. On the other hand, at the edge of the ice far off the coast of Thule, extensive banks lie in depths of from 130 to 265 feet. This area is notable for the presence of numerous stranded icebergs which, while no doubt destroying vast colonies of shellfish as they scrape across the bottom, do at the same time plough up the banks and release fresh "fertilizers" to encourage the growth of new communities. During the winter the tidal cracks around these bergs are often frequented by walruses, because they can maintain breathing holes in the thin ice that forms over the cracks: just as when the seaward edge of the land-fast ice extends across a bank they can swim far under the mile-wide stretch of thin, broken ice.

A herd may lie out on a floe near a suitable bank for hours or days at a time, intermittently diving for up to 10 minutes at a time, before surfacing for 3 or 4 minutes to take in air, or to sport around for a while before diving again. When a walrus dives to a bank he turns over, spreads his flippers and descends "upside-down" in the manner of a dive-bomber peeling off. He then, according to one observer, sinks to the bottom and, while "standing" on his head and moving backwards, ploughs up the bank with his decurved tusks, gouging out the mollusks if they are buried in gravel or boulder-clay, or probing for them if they are in thick beds of seaweed or among rocks. Such a technique would seem to be practicable if one takes into account the peculiar conformation of his tusks, the fact that they are invariably clean and smooth from use, and that they may ultimately be worn down to half their normal length. But since their tips are invariably heavily abraded on the front and sides, but not on the back, it seems more probable that they are rotated from side to side, rather than scraped backwards. The degree of wear varies considerably from one individual to another, depending upon whether a particular walrus has been feeding on a sandy or a rocky bottom. The tips are often broken off against rocks, and a whole tusk is frequently lost when perhaps a walrus, desperate for air, is making strenuous efforts to free it from a crevice. But since the majority of walruses with broken tusks are bulls—aged bulls may lose both—one suspects that many breakages occur when an ugli of bulls is stampeded into a mass rush over the rocks to the sea. Once a walrus is mature, when five or six years old, the rate of growth of its tusks is exceeded by the loss from wear and tear, and they may gradually be worn down to lengths of less than 6 inches in old bulls, and cease to grow after about twenty-five years.

Presumably a pod of feeding walruses ploughs up the bottom indiscriminately, for there does not appear to be any method by which they could pinpoint buried mollusks. Even if the latter's siphons were exposed above the mud how could these be detected by sight at depths of 200 or 250 feet, particularly since a walrus' eyes are not apparently adapted to function in conditions of low light intensity? Possibly his mobile *vibrissae*, which are furnished with abundant nerve-endings at their bases, enable him to "feel" any mollusks that are not too deeply buried. He may possess as many as 400 of these strong plastic-like "whiskers." Varying from a match-stick to a raven's quill in thickness, they bristle from thick fleshy pads on his upper lip, and are longest and most closely set at the corners of his mouth, where they may grow to a length of 4 or 5 inches and form a stiff stubbly beard as broad as a man's hand. But, whatever may be the method by which a walrus locates shellfish, how does he gather up and consume the vast quantity he collects? He has only an absolute maximum of 10 minutes in which to dive, locate the shellfish, excavate, collect and consume them, and surface for air again, spurting out water in streaming cascades; and he may submerge for no more than 3 minutes. According to Russian observers, he gathers together the mollusks he has excavated by rolling them up into a mass with his *vibrissae*. These are certainly subjected to noticeable wear and tear as he ages, and those of very old walruses may fall out, exposing the rough and wrinkled skin. The central ones in particular may be worn down to short stumps when the lateral ones, less exposed to friction with the shells or bottom, are still 4 inches long; and it is significant that the vibrissae of captive walruses, lacking the opportunity to grub out shellfish, may reach an abnormal length of almost 12 inches. All observers agree that the mollusks are consumed before the walrus surfaces and that only their soft parts are found in his stomach, which may contain from 50 to 100 pounds of their feet and siphon, representing between 1,000 and

3,000 clams! These shelled clams have indeed always been a favorite delicacy of Eskimos, who open up the stomach of a walrus as soon as they have killed it and, after rinsing the clams in sea-water, eat them in the raw state.

The problem is—how does the walrus dispose of the shells? According to Freuchen, after he has scooped up with his flippers as many of the mollusks as he can grasp he heads for the surface, leaving in his wake a trail of dirty water as the shells' matrix of mud is washed away; but shortly before reaching the surface he rubs the rough palms of his flippers together and, in so doing, squashes the shells. These then sink rapidly to the bottom, while the soft parts, having almost the same specific gravity as water, remain suspended or sink only very slowly. At this juncture a walrus can be observed bending backward to either side and moving his head very quickly as he sucks in the shelled mollusks. But rather than crushing the shells with his flippers, it would seem that the Thule Eskimos are more likely to be correct in asserting that the walrus *sucks* out the soft parts, sorting out the fragments of shell with his tongue and spitting them out. In support of this assertion we know that a walrus' mouth and tongue do in fact possess remarkable powers of suction. A young captive walrus has been seen to fracture the shell of a mollusk by a sharp blow from its vibrissae pads, and then suck out the soft parts, while another captive sucked holes in the wooden boarding of its pen in quite a short time. Moreover the animal breeder, Hagenbeck, noted that his walruses had no difficulty in sucking out the 5-pound metal drainage plug from their pool when it was filled with 3½ feet of water, and he also observed that they were constantly sucking at food or any other object on the ground, and juggling it around with their tongues.

Conservation during the 1960s has resulted in the Pacific herds extending their range in the Soviet Arctic, but the catastrophic decline in the overall number of walruses during the past hundred years must have opened up large areas of the Arctic to other potential shellfish-eating mammals; yet none has taken advantage of this vacuum. As we have already noted, a single walrus may collect in one day's fishing the feet and siphons of more than 3,000 clams. An average day's consumption by a herd of 2,000 walruses—a small one by comparison with those of last century—would amount to some 2 million mollusks. Undoubtedly the seasonal and annual variations in the regional distribution of the ice-fields oblige the herds to deploy over immense fishing areas, lessening the possibility of them fishing out any one area of banks; but consider the vast toll of shellfish taken from traditional banks summer after summer, winter after winter, by hundreds of thousands of walruses, and the astounding molluskan fertility that can replenish such collossal losses. Since the herds are known to have exploited banks in the same localities for several hundred and very possibly several thousand years, it must be presumed that the regenerative powers of shellfish are equal to all demands by predators. Possibly those that survive are surprisingly long-lived, for it has been suggested that some giant clams and freshwater mussels live for a hundred years.

14: Polar Bears—Nomads of the Arctic

The lives of all the Arctic mammals we have studied thus far have been ordered by the seasonal growth and disintegration, advance and recession of the ice. Polar bears, though technically land mammals, are no less dependent upon the ice's vagaries and, with one foot on shore, one in the sea and two on ice, could be considered the link between sea and land in the Arctic. They pass much of the year among the pack-ice, and its southern limits coincide approximately with those of their geographical range because the ice is also the home of their main prey—seals. The almost total absence of bears on the coasts of Greenland and the Canadian Arctic islands and off the mainland of eastern Siberia, for example, is due to the fact that because the inshore ice on those coasts does not break up every summer and provide leads of open water, the seals are obliged to migrate to other areas where younger ice and open water are available. Conversely, where the pack is in constant motion, often breaking up even in winter, as is the case off Alaska and the deeply fiorded north-east coast of Greenland, the resulting temporary leads and polynias are peculiarly suited to the requirements of seals, and on these coasts bears are comparatively more numerous than in most other parts of the Arctic. This is confirmed by Stefansson, who has described in *The Friendly Arctic* how:

> If the coast is open as in northern Alaska, you can go five or ten miles to seaward and find a place where the wind has broken the ice and where the cakes are in motion. Here you will find seals swimming about in the water like bathers in a pond, and the tracks of polar bears may meet you anywhere. But in places like Coronation Gulf there is land on every side, and the ice does not move from November, when it forms, until the following June or July, when it breaks up some two months after summer and green grass have come upon all surrounding lands. There are no polar bear tracks on this ice except in rare years.

The main haunts of polar bears are therefore the north coast of Alaska; within a quadrangle of the Canadian Arctic islands bounded by Devon Island and Baffin Bay in the north, and by Southampton Island and the Davis Strait in the south; off the east and especially the north-east coast of Greenland; among the Svalbard islands; and off Franz Josef Land, Novaya Zemlya, Svernaya Zemlya, the New Siberian Islands and Wrangel Island. Today, after more than a hundred years of incessant persecution, only about 10,000 polar bears still range around the Pole, and they are now so locally distributed that it is possible to fly 5,000 miles over the Alaskan ice-fields and not see one. On the other hand, when Wally Herbert and his companions of the British Trans-Arctic Expedition were nearing the end of their 3,620 miles polar crossing from Alaska late in May 1969, and were on the ice off Svalbard's North-East Land, they were encountering at least two polar bears every day from May 23 until June 10, when they were finally lifted off the ice by HMS *Endurance*'s helicopter. Wherever leads split up the broken ice hundreds of bear tracks extended in all directions. Although the bears made no active attempt to attack the members of the expedition, they were constantly interested in the dogs, and since neither shouting nor shots in the air deterred them, and they just came walking up, unheeding, Herbert felt obliged to shoot those that did not ultimately turn away at a distance of 20 feet. All animals in the Arctic are so accustomed to ice noise that they are rarely particularly alarmed by gun-shots. To quote Stefansson again:

> Ice floes crackle out at sea with a high-pitched screeching like the thousand-times magnified creaking of a rusty hinge, when one six-foot floe is forced over another by the remorseless drive of the pack, ever pressing on at twenty feet a minute. The floes buckle and bend, and are thrust up into great ridges; then tilt, snap and crash over, groaning and booming like surf cannonading on a rocky shore, as their edges are ground to powder. Bergs grind and growl. The pressure roars and rumbles like thunder, screams and howls; then dies away with a shrill whistling as of escaping steam, to a sudden stillness, before its guns boom out again.

Man and his possessions hold only one initial interest for those bears, particularly he-bears, bold enough to investigate—are they a source of food? All explorers and hunters have emphasized the polar bear's intense curiosity about them and their belongings. If a sledging party crosses a bear's path, he gets up on his hind legs the better to examine these strange creatures and sniffs with distended nostrils. He may even leap off the ground for a better view. Sledging parties are often followed for several miles, and the food caches of explorers, the dogs and refuse of Eskimo and Chukchi settlements, and especially the debris of whaling and sealing stations, must have significantly influenced the geographical distribution of polar bears during the past 300 years. A curiosity based on the possibility of humans being edible is natural, in view of the fact that the greater part of a polar bear's life is devoted to searching a white world for the dark figures of seals, or for the carcasses of whales or walruses cast up on the shore. But it is evident that to all except very hungry bears, man's own scent is not edibly attractive, and while stalking up to a sledge party he may lose interest on the way if he scents a seal. Some bears indeed will run away immediately they get the wind of a sledge party at a distance of 300 or 400 yards. These no doubt are bears that have learned to fear men. She-bears, whether alone or with cubs, are usually much less bold than he-bears; but the reactions of any polar bear towards man are unpredictable, though unprovoked attacks are not common.

It is probably true to say that there cannot be any stretch of coastline around the Polar Basin that has not been trodden at one time or another by polar bears. And there can be few sectors of the Arctic ocean whose ice-fields have not been traversed by them, though their visits to the North Pole and the Pole of Inaccessibility must be relatively infrequent, for in those regions the pressure in the pack-ice is at its maximum, piling up a chaotic jumble of ridges and hummocks lacking any extensive fields of level ice, while only occasionally does a temporary slackening of the pressure allow the floes to draw apart and open up a lead. Nevertheless, there have been a number of encounters with bears north of 80° on what may be termed the Soviet side of the Pole. A Russian expedition, drifting south-west of the Pole on an ice-floe station, encountered a she-bear with two cubs on August 1 in latitude 88, nearly 350 miles from the nearest land—northern Greenland. As late in the season as December another she-bear with a yearling cub became entangled in the wires of the runway lights at the U.S. air-base on the drifting ice-floe station Alpha, when the latter was at 84° N! Peary sighted fresh bear tracks along the edge of a lead when north of 86° on his way to the Pole from Ellesmere Land late in March. A he-bear was killed near Nansen's *Fram*, when she was frozen in at 84° north of Severnaya Zemlya, while he himself saw tracks north of 83° when sledging south to Franz Josef Land from his abortive attempt on the Pole. And, most recently, Wally Herbert shot a bear at 82° 27′ N when 500 miles from the Pole and 300 miles from Henrietta Island, the most northerly of the New Siberian islands.

These encounters illustrate the fact that polar bears are essentially nomads, hunting up and down the fiords and along the coastal ice-foot, rafting on floes from one island to another in the course of extensive sea voyages with the pack-ice and often traveling overland. Indeed, until quite recently, it was generally assumed that they drifted passively with the pack-ice from one part of the Arctic to another, habitually circumnavigating the Pole; and it is true that an individual's wanderings with the pack must involve him in voyages of many hundreds of miles. Witness the regular New Year arrival of bears at the breeding grounds of the harp and hooded seals on the ice-fields off Newfoundland. Although these bears could have started out from the south of Greenland more than 500 miles distant, it is more probable that they have in fact voyaged 1,000 miles or so on the stream of floes drifting south from Baffin Bay, for bears have greatly decreased in southern Greenland since the 1930s. They are also regular visitors to the seal rookeries on the ice-fields north of Jan Mayen, 300 miles from Greenland and 450 miles

Polar bears—mother and cub

from Svalbard, as they are to Bear Island when the drift of the pack is favorable from Svalbard 150 miles distant. When the spring ice extends unusually far south, or when great masses of ice break away from the glaciers of north Greenland and drift south, bears reach the north coast of Iceland. That, for most if not all of these bears, is journey's end. Ravenously hungry on arrival, but with no seals or carrion awaiting them, they move inland to prey on sheep, cattle and horses, and break into farm-steadings. If they are not killed by farmers when they first come ashore, they are accounted for later in the summer or autumn when they return to the coast and find that, with all the ice gone, they are prisoners on the island.

Freuchen stated that one could be certain of meeting bears far out to sea in those latitudes where pack-ice was regularly drifted by currents; and that such bears were usually fat because of the abundance of seals. The main streams of the pack may be said, with some exaggeration, to serve as moving highways on which bears (and also Arctic foxes) are transported in numbers; and as many as twenty-nine bears have been sighted adrift on a single floe. Freuchen added, however, that any bears drifting too far out to sea would leave the ice and swim back to land; and there have been many encounters with bears out of sight of land and scores of miles from the nearest ice, and also swimming out to icebergs 15 or 20 miles off the land—presumably in expectation of finding seals on them.

Many bears, hunting seals far out in the pack or blown off-course during storms, will certainly not be able to return to their starting points, but will be stranded on islands and mainland shores hundreds of miles from these; and whether or not bears arrive at a certain island at their traditional season depends upon the annual variations in the extent and course of the drift-ice. When onshore winds drive the pack against the land-fast ice one can confidently expect numbers of bears to come ashore during the next few days. But that bears should regularly circumnavigate the Pole is quite another matter. She-bears, having to den-up with cubs at least once in three years, obviously could not, while mating rendezvous with he-bears would prove impossibly haphazard. Moreover such circumnavigators would be obliged to pass through extensive "famine" areas where immense fields of solid ice would contain no seals, nor any other source of food. It would be contrary to all known animal laws for polar bears to seek deliberately to break through such formidable ecological barriers. No animal is going to abandon good feeding grounds so long as conditions remain favorable. Contemporary naturalists, while not denying that bears travel long distances with the pack, take a more conservative view of the extent of their wanderings. As yet only a few hundred polar bears have been marked with tags, and too few recovered to throw any light on their migrations. Of 60 tagged in the Hudson Bay area, for example, only one, a she-bear, has been recovered, in the same locality; while of 103 tagged in the Svalbard archipelago all but one of the 15 subsequently recovered have also been shot in the same area. The one exception, however, had traveled at least 2,000 miles to south-west Greenland during the eighteen months between marking and recovery.

A bear is a strong and indefatigable swimmer. His huge paws, serving as powerful paddles, propel him at a rate of 180 yards a minute and enable him to "porpoise" in 10- or 12-foot leaps. He is also admirably equipped to withstand prolonged immersion (and also low temperatures), with limb-bones that are porous, marrow-less and full of oil, a skin spongy and also extremely oily, a subcutaneous layer of fat or blubber, which may be as much as 3 inches thick on his heavy haunches, and a thick coat composed of dense under-fur and oiled outer hair, off which the water runs very quickly when he emerges. Yet it is a most remarkable fact that in general polar bears, except when hunting seals, apparently take to the water only in emergencies, particularly when harried by hunters or aircraft. When a bear dives, he seldom plunges deeper than 3 or 6 feet, and rather than plunging into a lead too deep to wade he lets himself down cautiously into the water backwards. Confronted by a broad stretch of water during his wanderings with the pack, he prefers to make for the shore or for an island rather than swim across to the next ice-field, and when homing north in the autumn he will trek a long way overland rather than swim across quite narrow inlets barring his way. Even when inshore he fastidiously avoids wetting his feet, detouring around large pools on the ice; and in the autumn every pool may be surrounded by the tracks of bears, most of whom have carefully avoided crossing the new ice, probably because crystals with needle-sharp upward-pointing tips form on the bottom of freshwater pools. Is it their awareness of being totally defenceless, when swimming, against the attacks of killer whales and walruses that accounts for this reported aversion of most polar bears to venturing into the water?

Once a wandering bear has gone ashore from the pack it follows that, unless he goes out again with it before the floes break up, he will be stranded ashore for some months; and this is what often happens. In these circumstances she-bears may be obliged to den-up for the birth of their cubs on islands, while still en route to their traditional denning localities, and not be able to get off them again the following spring before the ice breaks up and floats out to sea. Other bears may have to return overland to their starting points. The ability to "home" over great distances is common to every animal except contemporary man. As bears age, they no doubt become acquainted with the physical geography and memorize landmarks over a considerable range, and there is evidence that they home on a direct bearing, although this may involve traversing necks of land, broad projecting spits, and even the entire breadth of peninsulas. Although the majority of polar bears rarely venture more than a mile or two in from the coastline, there is no reason to suppose that those encountered beyond the normal limits of their range have necessarily lost their bearings. One would expect such confirmed wanderers to turn up in improbable places. In Alaska, Canada and Siberia alike polar bears have been known to penetrate 100 miles or more inland. In the summer of 1962 fresh tracks of bears were found on Svalbard's 5,600-foot Mount Newton; and tracks have also been seen more than 30 miles in from the east coast and at a height of 5,000 feet on the most inhospitable land in the Arctic—the Greenland ice-cap, where the traveler is aware, as Robert Peary expressed it, of only three phenomena, "The infinite expanse of the frozen plain, the infinite dome of the cold blue sky, and the cold white sun; and where he may travel for days or weeks with no break whatever in the continuity of the sharp blue line of the horizon." A polar bear can undoubtedly fast for periods of days or even a week or two without losing strength, and David Haig-Thomas speculated in *Tracks in the Snow* on the possibility of an occasional bear actually crossing the ice-cap from coast to coast—a traverse varying from 200 to 700 miles; for the tracks of a thin and exhausted bear with all the skin worn off its paws, which he killed at the foot of a glacier, led straight across the cap for as far as he followed them.

For most of their adult lives, then, polar bears, in contrast to walruses, are solitary nomads. Wandering he-bears, indeed, will turn aside from their course rather than meet. Mature he-bears are tyrants whom young bears avoid, and with whom she-bears consort only for the few days, or possibly weeks, during the mating season in the spring and early summer and whom they shun when fearful for their cubs' safety. But in general polar bear society is one of "armed neutrality" in which unnecessary conflict is restricted to the mating season and to that competitive season of hunger in the spring when those bears that have denned up during the winter may, when they break out, have to travel long distances to the nearest seal waters. Although, according to Alwin Pedersen, no Greenland bear would have to trek for more than ten or twelve days from its denning place before finding seals, in regions where the latter are scarce bears must often be near starvation. If deaths occur in the spring they are usually those of younger or weaker bears who have been fortunate enough to secure a seal or carrion and, emboldened by hunger, show fight against stronger bears or are taken by surprise.

If seals are hard to come by during the winter, or a herd of belugas or narwhals has not been trapped in a polynia, those bears who have not denned up or who have not gone out to polynias in the pack-ice, must resort to scavenging in order to fill their bellies; and any object offering the slightest likelihood of being edible attracts them. Tree trunks, stranded on raised beaches, are rolled over and closely examined, and a bear's stomach may be full of wood-chips gnawed from driftwood; but the carcasses of whales, seals and walruses, and also those of other bears, are their main source of food during the hungry months. Even a fat bear often prefers putrid carrion to fresh seal blubber, and one may settle down for an entire winter beside a large carcass. The Danish zoologist, A. L. V. Manniche encountered a large and extremely fat he-bear who had been raking about in the carcass of a walrus for so long that the hair on his neck had rotted away, and he had hollowed out the cadaver to such an extent that he was standing within it, partly hidden by what skin remained, while devouring the internals. Engrossed in his meal, he took little notice either of Manniche or of the large numbers of attendant ravens, glaucous gulls and foxes. The carcass had also been visited by wolves, and in order to keep watch over it, yet be able to rest when not feeding, the bear had excavated a deep hole in the sand close by.

On those coasts where bears have drifted ashore with the pack the search for carrion may begin as early as August, and by the onset of winter numbers of bears are likely to assemble shortly after any marine mammal has been killed or cast ashore. There are several records of as many as a score of bears, together with upwards of a hundred foxes, congregating in the vicinity of a dead whale. As recently as the early 1950s an Eskimo sledge party counted forty-two bears in the immediate neighborhood of a bowhead stranded on the north-east coast of Southampton Island. In bygone days, when polar bears were numerous, they gathered at a carcass with uncanny speed; and in 1896 the captain of a ship, frozen in the pack near the stranded carcass of a whale east of MacKenzie on the Yukon coast, is reported to have shot thirty-five bears, one after the other, as they arrived at the carcass. Since polar bears are, as we have seen, essentially solitary nomads hunting separate beats, how are they attracted in such numbers and in so short a time to these carcasses, which occur only sporadically? Decomposition is very slow in polar regions, continuing into the second year after a large animal's death, and there is comparatively little odor because of the low temperatures prevailing in most months. Nevertheless, there can be no doubt that it is a bear's phenomenal scenting powers—which have impressed all Arctic men since the earliest whalers and sealers—that account for these assemblies. Although we may doubt the reputed ability of a bear to scent a carcass buried deep in a snow-drift at a distance of 10 or 20 miles, as experienced a trapper and naturalist as Pedersen vouches for its ability to wind carrion at a distance of more than 12 miles. We must also be impressed by its skill at locating the aglos (breathing holes) of seals, and it is the abundance of seals that has made it possible for this largest of all the carnivores to inhabit a polar environment and become a semi-marine animal.

Virtually all occupied aglos are so overlaid by snow as to be invisible to human eyes and, presumably, to those of bears also; while an uncovered hole in a patch of ice from which the wind has swept the snow usually proves to be unoccupied, either because the unprotected ice freezes too quickly for the seal's liking, or perhaps because the seal senses itself to be endangered by the exposure of the hole. But bears are certainly very interested in all holes in the ice, frequently investigating those cut by ships' crews when watering; and they locate the snow-covered aglos with such unfailing regularity, no matter how deep the overlying snow, that no polar inhabitant has ever been in any doubt that they scent them out. In some instances, however, when the ice is thin, they may perhaps be attracted by the water spurting up when a seal blows with a long bubbling gasp, like a jet of steam escaping, and which is audible half a mile away on a calm day.

Once a bear has discovered several aglos in use he makes regular rounds of them, though with every seal blowing at a number of holes he may have to wait beside one hole for a very long time before it is visited. First he scrapes away the snow and the uppermost layer of ice with his 3-inch claws, until he has enlarged the aglo sufficiently to insert his broad paw. Then he either sits back and wait patiently for a seal to blow, which it is likely to do every 7 or 8 minutes at one or other of the holes, or he lies some 3 feet away from the hole with head resting on outstretched paws—a position he can retain for long periods without moving a muscle. Ringed seals are not only sensitive to sound-waves traveling over the ice and through the water, but also have keen hearing. A bear, however, can approach noiselessly on his furred pads over the snow-crystals on the ice, and the weak link in a seal's defensive faculties is that it cannot hear at the actual moment it is blowing. Thus the instant its muzzle appears in the breathing aperture the waiting bear strikes its head with his claws and drags its body out on to the ice almost in one movement. His neck, shoulder and forearm muscles are so enormously developed that he is powerful enough to drag a bearded seal, almost his own weight, by main force out of its hole, crushing its shoulders, ribs and pelvis and squeezing out its intestines in so doing. However, it is not unknown for a bear, in striking at a seal, to plunge head-first into the aglo and become trapped by the neck in the narrow opening.

The polar "night" begins in the late autumn when streams freeze in the fiords and when only the midday sun is strong enough to melt the new ice, until this too freezes permanently over the bays. With the coming of the first heavy falls of snow shortly after the middle of October the sun rises for the last time in the High Arctic, though it will be seen for another month in lower latitudes:

> Today [logged Nansen on 26 October, when off the New Siberian Islands] we took solemn farewell of the sun. Half of its disk showed at noon for the last time above the edges of the ice in the south, a flattened body, with a dull red glow,

> but no heat. To console us for the loss of the sun, we have the most wonderful moonlight; the moon goes round the sky night and day, over the great stretch of white, shining ice. And in the midst of this silent, silvery ice-world, the northern lights flashing in matchless power and beauty over the sky in all the colors of the rainbow.

By the first week of November in the High Arctic, towards the end of the month further south, midday is indistinguishable from midnight except for a faint twilight for a few minutes at noon when the sky is clear. However, even on an overcast night the stars behind the pall of cloud reflect sufficient light from the snow to reveal a man in dark clothing at distances of from 10 to 50 feet. By starlight the ring of freezing breath around a raven's neck is visible some yards away, and at midday, when the moon is not obscured, the bird's tracks in the snow. On clear moonlight nights a bear can be plainly seen half a mile away. Above leads of open water, freezing vapor rises in dense black clouds like the smoke of prairie fires, as the temperature falls to 30 and 40 and ultimately to 60 or 70 degrees below zero, and the "urine runs upwards" observes the Eskimo as his water freezes before reaching the snow and becomes an inverted icicle.

As the winter freeze-up extends, so it becomes increasingly difficult for bears to locate polynias of seals, and long vigils at their aglos are not always rewarded. Thus although many bears remain in good condition throughout the winter, the bellies of others must be empty more often than not, and it is probable that bears of all ages den-up for variable periods during blizzards and very low temperatures. A he-bear may do no more than curl up in the snow and sleep the hours away when a severe blizzard renders any movement impossible, or dig out a temporary hole as a shelter or merely as a wind-break in which to sleep off a heavy meal. But some he-bears and barren she-bears lie up in small holes in the snow on shore, or out among the pressure-ice, for a week or two at a time, and a minority of these den-up for several months.

The presence of an inexhaustible supply of food in the form of seals was not in itself sufficient to make life possible for bears in the Arctic. With temperatures falling as low as −70 or −80 degrees F the young of a large carnivore, slow in maturing—a polar bear's cubs are no larger than rats or guinea-pigs at birth—could not survive the long months of winter in this frozen world, devoid of all natural shelter, without some special provision. This has been met by the ability of bears in general to go into partial hiberation, and by that of polar bears to excavate dens in the snow. And it is in October or November, or as early as the middle of September in certain localities such as Wrangel Island, that the older pregnant she-bears leave the pack-ice in order to prospect for suitable denning sites on the land-fast ice or on the land itself. Younger she-bears may delay denning-up. Not all she-bears are able to make a landfall at the traditional denning places of their particular geographical group because they are, as we have seen, to some extent helpless drifters with the pack, and may be carried hundreds of miles from their starting points. Alternatively, since winds are as influential as tides and currents in controling the movements of the pack when it breaks up, untimely directional changes in the wind may result in the pack and its floe-rafting bears not reaching a traditional denning area in a particular year. Nevertheless, the normal directional drift of the various polar currents combines with local snow conditions to render certain regions especially suitable for denning she-bears—notably north-west and east Greenland, the eastern islands of Svalbard, Franz Josef Land, Wrangel Island and, in the Canadian Arctic, the southern part of Banks Island, the Simpson Peninsula, and the eastern sectors of Southampton Island and Baffin Island.

In favorable years there may be considerable concentrations of she-bears in these traditional denning areas, and the same dens are occupied year after year, though not necessarily always by the same individuals. The Russian zoologist, S. M. Uspensky, traversing more than 900 miles of Wrangel Island by dog-sledge and snow-cat during March and April 1964, located with the assistance of tracker dogs 116 out of a probable 150 or more dens on the island, some in groups of two or three to the square kilometer. Since a pregnant she-bear's den is to be her permanent home for several months, the temporary hole in the snow of the average he-bear is not suitable. She requires special snow conditions, and although half the dens on Wrangel Island and more than 60 per cent of those in the Canadian Arctic are situated less than 5 miles from the sea, and none as far as 20 miles from it on the former or more than 30 miles in the latter, some she-bears travel considerable distances inland in search of suitable denning places. In the Ontario sector of Hudson Bay, for example, their dens have been found by Indians in spruce muskegs more than 100 miles inland; and since polar bears and brown bears are closely related it is interesting to note that both their dens in the muskegs are excavated in similar sites—a cavity formed by the upturned roots of a fallen tree, or the underhung mat of vegetation on the bank of a river. Some she-bears den-up in sheltered places among the hummocks of pressure-ice and in exceptional instances out in the pack, on floes and bergs in fiords and on glaciers. In Greenland they seldom den-up offshore, but half a mile or more inland, while on Svalbard most retreat to the mountains where snow-drifts form early in the autumn, and may ultimately dig in at heights of almost 2,000 feet.

What special conditions does a she-bear require? In the first place the density of the snow is very important, because dens cannot be excavated in snow that is very hard or very soft. Next, the snow must lie in deep drifts, as it normally does on the lee slopes of coastal hills and valleys; but since an excessively heavy covering of snow would render it difficult for a bear to dig her way out in the spring, most dens are excavated on the upper third of the hill slopes—where gradients of from 25 to 40 degrees will hold 7 or 10 feet of snow—and very few at the foot of the hills. However, interesting regional differences occur in the choice of sites, for whereas more than 75 per cent of the dens on Wrangel Island are located on the northern and eastern slopes which receive the least sunshine during the summer and therefore retain much of the previous winter's snow, 75 per cent of those in the Canadian Arctic are on south-facing slopes, where the cubs can exercise and bask in the spring sunshine.

During her prospecting for a suitable site a she-bear wanders around for a long time before finally selecting a hollow or cavern in a drift of old snow and settling down to enlarge it. As she excavates, so her point of entry is filled with the snow she scratches out behind her, and blizzards complete its sealing off and also cover any tracks that might betray the den's whereabouts. A number of occupied dens have been examined, particularly on Wrangel Island and in the Canadian Arctic, and as a general rule a typical den is entered by a passage from 6 to 9 feet in length and not less than 2 feet high. This slopes slightly upwards to a low ridge, beyond which are either one or two, or exceptionally as many as four, oval-shaped chambers from 6 to 8½ feet long, 4 to 5 feet wide, and 3 to 5 feet high, with walls and ceilings firmly packed and scored with claw-marks. A polar bear's den is clearly the prototype of the Eskimo's igloo (or snow-house). In dens with two chambers, the second and larger one is probably excavated by the she-bear as a birth-place for the cubs. However, considerable variations are possible. A Southampton Island den opened up by Richard Harington, the Canadian zoologist, measured 10 feet by 8 feet and was 4 feet high, while a Svalbard one was divided into two chambers, connected by a narrow passage, each 10 feet long and 5 feet wide.

The she-bear's five or six months' vigil within her den, during which period she subsists solely on her reserves of fat—though she can slake her thirst from the walls of the den—is often alluded to as a hibernation. But though her body temperature may be a degree or two lower, her heart-beat rate a little slower than normal, and her condition lethargic, she does not relapse into that state of torpidity associated with true hibernation, for she remains in full possession of her senses, not only giving birth to her young but wakening instantly if disturbed. When, for example, on a very cold February afternoon. Harington punched a hole in the roof of a Southampton Island den, in order to ascertain the temperature of the interior, a glistening black eye and twitching muzzle were immediately applied to the aperture; and the she-bear subsequently paced around uttering peevish grunts, while her two cubs huddled against the far wall of the chamber. (Although the she-bear maintains an air-hole from 3 to 12 inches in diameter in the roof of the den the interior temperature may be as much as 40 degrees higher than the outside air temperature.) Nevertheless, if a she-bear does not hibernate she must, in order to conserve heat, energy and fat, sleep away the greater part of this long retreat; but she sleeps lightly, "sitting" back against the wall of the den, and wakening at frequent intervals to suckle the cubs, which cling, warm and purring, with their curved needle-sharp claws to the soft fur between her thighs, interposed between them and the frozen floor of the den. Moreover, if she is disturbed by an avalanche, or by the movement of a berg if she has denned-up on the ice, she will break out of the den and carry her cubs in her mouth one at a time to another shelter.

The cubs, usually two in number, exceptionally three, very rarely four, but often one in the case of younger she-bears, are born mainly in December and January. What determines the end of the nursing she-bear's denning vigil? Is she aroused to break out by increasing light percolating through the snow walls of the den for a longer period each day in March? Unlikely, in view of the fact that most dens are buried beneath 1 to 7 feet of hard-packed snow, though possibly a little light filters into those dens with large air-vents and illuminates their interiors with a soft bluish light on a sunny day. Moreover a few bears are apparently in the habit of breaking out periodically as early as February, although the sun does not rise above the horizon until late in January in the Low Arctic and a month later in the High Arctic. Possibly hunger drives her out, though she will eat very little at first, taking only berries and grass preserved beneath the snow until she is in a fit condition to begin hunting seals.

The cubs themselves cannot be allowed out until they are

well furred, mobile and strong enough to survive the fatigues and rigors of the polar spring when storms, though short-lived, can be more severe than at any time during the winter; when temperatures are still falling as low as −50 or −60 degrees F; and when the sun's heat is still not strong enough to melt the snow. If they are fortunate they will come abroad for a first glimpse of their white polar world on a bright sunny day of no wind—some in mid-March, but the majority not until April—and for the next three to five days, or longer if there is a snow storm, the family makes only short excursions in the immediate vicinity of the den, returning to sleep in it at night or to shelter in bad weather. These initial all-important days are mainly devoted to accustoming the clubs to their new environment and strengthening them by various forms of play, with the elevation at the outer end of the den's passage serving as a practice climbing pitch and tobogganing slide. Later, their play can be extended to sliding on their bellies down the nearest ice slope with legs outstretched fore and aft, in imitation perhaps of their mother sliding with them; and Alwin Pedersen watched one she-bear, stationed at the bottom of a slide, catching the cubs with her paws every time they arrived at the bottom, plastered with snow.

From their very first day out of the den the cubs display remarkable powers of endurance, following in their mother's footsteps courageously, one behind the other, though they keep at her side in soft snow, in order not to fall into the deep holes made by her huge feet, almost 3 feet in circumference, for they stand only between 9 and 12 inches at the shoulder. However, she does not allow them to become overtired, pushing them along with her muzzle between their hind legs, or carrying them on her back. When the den is situated some distance inland the cubs may be taken direct to the sea-ice in the fiords or on the coast, but during their first week of travel the she-bear makes frequent stops to play with them—ducking them in leads, sliding down the sides of small bergs, tumbling and rolling them and making pretence at fight, or allowing them to suckle and sleep between her legs. For the latter purpose she selects a block of ice or a large rock, in the lee of which they can recline, sheltered from the freezing wind, and bask in the sunshine glittering on the snow-fields and hummocks of ice. From time to time she climbs one of these hummocks in order to survey the surrounding country, and if she spots a he-bear or wolf or man leads the cubs away on a circuitous detour, or hides them among the hummocks. Whenever possible she avoids the land, where the cubs are in more danger from wolves when she is away hunting and has left them hidden behind a rock or hummock: whereas out on the sea-ice there is only sporadic danger from wandering he-bears or, exceptionally, a bull walrus.

If the winter has been severe and prolonged it may be March before a he-bear can get himself out of his snow shelter and lie in the sun against the broken wall of his den. In other years fresh tracks of bears, leading to fiords and coast, may be seen early in February, while bears who have wintered out on the pack may also come inshore at this time. All these early and ravenously hungry he-bears are bound for the same goal as the she-bears with cubs will be subsequently, for as we have seen, it is at this season, providentially for bears, that the ubiquitous ringed seals give birth to their pups on the ice. From now on through the spring and early summer the bears will enjoy their best hunting. Once again their phenomenal scenting powers enable them to find these birth aglos regularly, despite the fact that they are almost invisible; and only those pups in aglos covered by more than 3 feet of snow are safe from detection.

Once she has led her family out on to the ice the she-bear is mainly concerned with this search for seal pups for herself and her cubs. Before setting out on a hunting sortie she settles the cubs in a safe place some way in from the ice, having first assured herself that there are no potential predators in the neighborhood. And there they remain until she returns. Later, when they are strong enough to accompany her on hunting expeditions, they lie down to sleep or play while she sets about breaking into an aglo. However, not all cubs are amenable to discipline, and Freuchen has described how some hunting she-bears are seriously hampered by them: "Time after time they lose patience when matters go too slowly and will begin to move about," he wrote, "then the mother will steal back and with violent blows of her paws persuade them to remain still, after which she starts her stalking all over again. Naturally this will often be repeated, and very often when she is so close to the seal that it scents the danger and puts an end to the hunt by diving into the water."

If successful in locating an occupied aglo the she-bear excavates most of the overlying dome of snow very rapidly with blows from her paws, then, rearing up, she pauses for an instant to estimate the distance, before plunging down with the full weight of her body on her fore-legs. Alternatively, she may retreat a little way from it, and then trot up and jump on the dome. She may have to do this again and again before finally breaking through the hard-frozen crust and, thrusting in her paw, hook up the pup with her claws.

Once the seal pups have abandoned their aglos the bears

must hunt along the leads and ice-cracks, and throughout the early summer many families stay out on the ice, seldom visiting land. Because young seals surface to breathe three times as often as adults, they are correspondingly easier to capture. Moreover, these "silver-jars" (as they are known after they have shed their white baby-fur), inexperienced in the ways of bears, may bask on the ice near hummocks affording cover to a bear or near a hole in the snow in which a bear has concealed itself with only its head erected to command a clear view over the frozen sound. Or again, after rolling over in the snow a few times, the pups forget where their holes are and, pulling themselves along with their flippers, stray far across the ice-fields—easy prey for bears or foxes or wolves. When young seals are to be had for the taking, a bear becomes a delicate eater, merely crushing a pup's head or taking a few mouthfuls of skin and blubber, before abandoning the carcass and killing another pup.

Adult seals, on the other hand, are extremely difficult to surprise, having the sense to lie out on level expanses of ice, well away from any hummocks or land, and are continually on the *qui vive* in regions where bears are numerous. All lie with their heads facing away from the wind, maintaining a constant look-out to leeward and, when lying close together, take turn about in keeping watch:

> A seal does not crawl unguardedly out on the ice; he is always fearful of polar bears [observed Stefansson]. When he wants to come and bask, he spies out the situation by bobbing up from the water as high as he can, lifting his head a foot or two above the general ice-level. This he does at intervals for some time, perhaps for hours, until he concludes that there are no bears around and ventures to hitch himself out on the ice.
>
> Here follows another period of extreme vigilance during

> which the seal lies beside his hole ready to dive in again at the slightest alarm. Eventually, however, he begins to take the naps that were his desire in coming out of the water. But his sleep is restless through fear of bears. He takes naps of thirty or forty or fifty seconds or perhaps a minute. Then he raises his head ten or eighteen inches from the ice and spends five to twenty seconds in making a complete survey of the horizon before taking another nap.

Only far out in the pack and in regions where bears are now uncommon, such as southern Greenland, will you see an adult seal rolling about in the snow for a quarter of an hour, before stretching out with its head on the ice and sleeping for five or ten minutes at a time. However, when they are molting in the summer both ringed and bearded seals become lethargic and less watchful, and though some take the precaution of hauling out on isolated floes for this purpose, most are easy prey in this condition. Some indeed sleep so soundly that a man can walk right up to them, and Stefansson cites an instance of a wolf capturing a seal asleep beside its hole and killing it by biting its throat. But normally a bear is obliged to take infinite precautions when setting out to stalk a seal, moving up-wind, taking advantage of any ruggedness on the ice, and sinking on to his belly when the seal is still only a black blob on the ice. No doubt the off-white color of a bear's winter pelage has some camouflagic value, while his yellower summer coat tones well with the floes of old yellowish ice floating in the broken summer pack. Pulling himself along on his belly over the ice, hind legs trailing, a few yards at a time, with head and neck thrust forward only an inch or two above the ice (or pushing himself along with his hind paws, with forearms doubled under his chest) the bear creeps up yard by yard until within 12 or 15 feet of the seal when, if it has still not been alarmed, he launches himself at it in one or more lightning bounds and strikes it with his left paw—invariably apparently with the left one. Then, after licking up the blood gushing from the wounds, he pulls the carcass 3 or 4 feet back from the edge of the ice to prevent the possibility of it slipping back into the water. From 10 to 40 pounds of the blubber usually satisfies him, and the bulk of the flesh is left untouched for foxes, gulls and ravens, though he is capable of consuming an entire seal.

Later in the season, when the veneer of snow over the ice has melted and the seals' holes have been enlarged, a hunting bear will take advantage of this change. Chancing upon such a hole 40 or 50 yards from a seal basking beside its aglo, he "freezes" until the latter's head sinks for a brief nap and then slips down the hole into the water. Swimming upside down beneath the ice he makes directly for the aglo, guided perhaps by the sun shining down it, and waits for the seal to dive into his paws.

Hunting seals among the broken pack-ice demands a different technique, dependent on the pack being provided with hunting platforms and stalking cover in the form of pressure-ridges and hummocks, together with broad leads or polynias where seals can surface in numbers. Perceiving a seal basking on a floe several hundred yards distant, the bear stands erect for a few seconds in order to take his bearings —as he frequently does when hunting, often climbing a hummock in order to obtain a wider field for eyes or nose. Then, going to leeward of the seal, he approaches the edge of the lead and, digging his claws into the ice, lowers first one hind leg and then the other, and backs down carefully into the water, until only his muzzle remains above the surface. Now, with small ears flattened back to wedge-shaped head, and only his nose and intermittently his eyes above water, he swims smoothly towards the seal's floe. The only movement that can betray him is the very slight ripple made by his nose; and if the seal's suspicions are aroused by this he submerges and swims underwater. When some ten yards from the floe he raises his head cautiously for a re-appraisal of the situation, then dives again and, if he has estimated the seal's position accurately, he shoots up at the edge of the floe, shattering the skin of thin ice around it with his head, and strikes the surprised seal almost instantaneously as he surfaces. Should the seal be lying too far in from the edge of the floe to be struck from the water, the bear swings himself swiftly up on to the far side of the floe and hastens to cut off the seal's retreat to the nearest water, in which his superior agility on the ice is usually the deciding factor.

By July the ice has drifted out from most bays and fiords, coastal waters are open, and shoals of tom-cod run up the fiords, harried by seals and narwhals. Most bear cubs, having molted into their coarser adult coats, are weaned and have acquired a taste for seal's blood and blubber, though some will be suckled for a further eight or nine months. On the other hand, many of the he-bears and barren she-bears, wandering with the pack, land on islands and mainland shores for a change of diet, becoming partial vegetarians. If meadows are not the typical haunts of polar bears, many have browsed in them, though mosquitoes probably discourage extensive journeys inland. Grasses, sedges, lichens, mosses, sorrels and berries are all eaten with relish by summering bears; and especially the profuse crops of crowberries, cranberries and bilberries (the blueberry of North America) which ripen in August and are quickly frosted, to be preserved in their pristine freshness under the snow until the following May or June. If there is an abundance of berries a bear may eat no other food for weeks at a time, glutting himself to such an extent that his muzzle and hindquarters become stained blue with bilberries. On offshore islands bears are often to be seen eating large quantities of grass and other herbage, whole squares of turf being torn up; and the large numbers of bears that formerly drifted with the ice through Bering Strait in the late autumn regularly wintered on the bigger islands that lay directly in the path of the pack. Many of these remained on the islands when the ice retreated north in the spring and, after moving to the higher ground, stayed the summer there eating grass, scavenging carrion, and digging puffins out of their burrows. A vegetarian diet may indeed be a necessity after a winter and spring's blubber and carrion eating, though on Svalbard quantities of seaweed are eaten in winter and summer alike, and a bear will dive repeatedly to bring up large tangles and eat the choicest of them. Grass and berries can be supplemented with cast-up marine animals, mussels, starfish, shrimps and crustacea; but though polar bears feed on salmon in some localities, scooping them out of the spawning redds with lightning thrusts of their paws, or driving them into ice-cracks in river mouths, the extent of their salmon fishing is very limited and cannot be compared with that of the Alaskan brown bears, for whom salmon constitutes the main food in summer, and whose geographical range coincides closely with that of the spawning rivers. It is interesting to note that as soon as the salmon have finished spawning and become unfit to eat, these Alaskan bears return to a diet of grass and berries until it is time for them to go into hibernation.

A polar bear's other prey on land includes foxes and hares which, though usually too nimble for him to catch, are removed from traps, and he will also go the rounds of one trap after another as soon as they are baited with ptarmigan. An occasional ptarmigan may be caught when luring a bear away from its chicks with fluttering flight. When lemmings are numerous, numbers of these aggressive little rodents are captured by bears turning over the stones above their runs and burrows and, if they are she-bears, thrown to their cubs. Dead kittiwakes and guillemots are scavenged at the base of the vast loomeries on the cliffs, and eggs are plundered from the nests of duck and geese and tossed around playfully when the marauder is satiated. Bears and reindeer do not meet very often, except on Svalbard, for the deer pasture in the ice-free valleys during the summer and on

Arctic hare

the lichen-clad hills in the winter, going down to the coasts only in the late autumn in order to feed on seaweed; and there do not appear to have been any instances of bears attacking either caribou or reindeer, although they have been seen grazing the same meadow. Nor can encounters between bears and musk-oxen be common, and although there are a few records of these oxen being killed by bears, their herd defence is so efficient that even a large he-bear could probably kill only a solitary aged bull or a cow with her calf.

By the early autumn most bears are out on the pack-ice hunting seals again. During the spring and summer the cubs have grown very rapidly and now measure more than 4 feet in length and weigh around 130 pounds. Polar bears appear to be extraordinarily successful in rearing their cubs to maturity, and most of the thousands of sight records refer at all seasons to she-bears being accompanied by two cubs. Opinions as to when the family breaks up are conflicting. Some she-bears are reported to desert their cubs as early as their second spring, when they are fifteen or sixteen months old, immediately before or actually at the time she mates again. But some cubs are still being suckled at this age, and some families remain intact throughout another summer and possibly den-up together for a third winter.

15: The Nature of the Tundra

In the preceding chapters we have been mainly concerned with Arctic seas, but in fact the predominant physical feature of the region is the tundra. This has no counterpart in the Antarctic—unless one considers the tussac country to be tundra—because the tapering land-masses of the southern hemisphere terminate in latitudes too low for the formation of this type of terrain.

The tundra, the Finnish *tunturi* or tree-less plain, stretches for 4,000 miles from Bering Strait to the White Sea and, as the barren-grounds of North America, embraces an arc of a million square miles from the MacKenzie delta in the west to Hudson Bay in the east. A vast emptiness of snow in the winter; the tundra is an endless rolling "desert" stippled with shallow lakes and transected by streams in the summer, when the ground thaws out for a few weeks to a depth of inches or feet above the permafrost. It is a land of low heath and extensive bogs, of dry windswept tracts with only scattered herbage or merely lichens on outcrops of rock and boulder, though with oases of richer vegetation in the driest places where the snow melts earliest and where drainage is superior. A peculiar form of desert, with temperatures ranging from −70 degrees F in the winter to +70 degrees F at high noon in the summer. Rainfall is low, yet because there is little evaporation there is an abundance of water, unable to escape down through the granite-hard bed of permafrost, varying in thickness from a few yards to considerably more than 1,000 feet. Yet, again, because the water is in a frozen state, it remains inaccessible to plants and animals for the greater part of the year. Nevertheless it is the short cool summer, rather than the length and severity of the winter, that determines the character of the flora and avifauna. Plants can only flower during the brief summer thaw, and even then are liable to be desiccated by strong winds or smothered under snow; while the thaw, and its associated plague of myriads of insects, transforms the tundra into the perfect breeding grounds for tens of millions of wading birds and waterfowl. It is a macabre thought—though probably no more—that, should the permafrost line continue to retreat north at its present rate of perhaps 1,000 yards a year, under the influence of higher winter temperatures, a time might come when all the waters would drain away from the tundra, which would degenerate into a true desert in which no waders or duck or geese could nest.

With the exception of such polar deserts as the Greenland ice-cap and the inland ice on some of the High Arctic islands, Arctic lands possess, as we have noted, vastly more vegetation and therefore insects, than Antarctica and its islands. They can thus support a terrestrial herbivorous and carnivorous fauna, including some birds and mammals which are entirely independent of the sea. The longer and milder winters of the Low Arctic indeed encourage a continuous mat of such dwarf vegetation as heaths, shrubs and willows in some regions, providing grazing for sheep in southern Greenland. There are some 500 species of flowering plants and ferns in Greenland, of which as many as 90 grow in the north-east "desert" of Peary Land within 500 miles of the North Pole. There, and on such Canadian islands as Melville and Banks, both north of 70°, the meadows near the rivers are bright with a profusion of flowers—flax and white saxifrage, the small white roses of the mountain avens, and the white stars of stitchworth. There are green patches of dwarf dandelions, yellow carpets of cinquefoil, and fields of red and golden poppies. Purple saxifrage, cotton-grass and ferns grow between the deep banks of drifted snow, and slopes of lush blue-grass, reaching down to the edges of glaciers, attract the musk-oxen down from the screes and moraines to browse in those flowery green "temperate" meadows where bees hum and birds pipe. Otto Sverdrup, the captain of Nansen's *Fram*, continually referred to the abundant summer grass, "growing in waving patches for great stretches, in which there were multitudes of red and yellow, sweet-scented flowers," on Ellesmere Island and Axel Heiberg Island.

In August the tundra is green with mosses and some hills are ablaze with purple, yellow and white heaths and red, yellow, orange and gray-green lichens. The latter include the beautiful yellow-orange *Caloplaca elegans*, whose growth is stimulated by the nitrogenous droppings of cock ptarmigan on their special look-out boulders and by those of sea-birds on their nesting cliffs, whether these be on Svalbard, the fulmar-haunted bastions of West Greenland or the mighty buttresses of eastern Baffin Land, where Cape Searle may hold the largest of all fulmar colonies. The acid soil of the Arctic lacks nitrogen, and such plants as scurvy-grass, chickweeds, poppies, buttercups, brook saxifrages and golden saxifrages, alpine foxtail and other grasses and rushes thrive, especially in meadows and on cliff sidings fertilized by guano from the bird colonies. All Arctic vegetation survives and colonizes only by constantly adapting to its harsh environment and the acid soil. Almost all plants are small and stunted and grow in compact cushions, rosettes or tufts. The branches of shrubs tend to trail along the ground, and the stem of a creeping willow or juniper or Lapland rhododendron, no thicker than one's thumb, may show 400 annual rings of growth. All are adapted to resist desiccation and abrasion by the wind or by drifting snow or gravel; while as a defence against prolonged drought their leaves are often leathery and rather small, and both leaves and stems may be covered with a dense mat of hairs which lessens evaporation.

They also adapt their reproductive mechanism to the scarcity of insects, of which the whole of Greenland can muster only about 600 species. Pollinating bumble-bees do their best, being active in weather too cold and windy for other insects—perhaps because the compact and hairy nature of their bodies enables them to conserve the heat generated by their rapid wing-beats; but the inadequate supply of insects has obliged large numbers of such plants as bilberry, alpine bearberry, white heather and the blue mountain heath to be self-pollinating. Although the majority of species have abundant seeds these do not germinate in the cold soil prevailing from midsummer to autumn, and must wait to do so until the snow melts in the spring. Even then low temperatures result in most dying without germinating. Only the few, such as the ice-buttercup, which is the first plant to colonize the gravel exposed by a retreating glacier, can survive several years' burial beneath semi-permanent snowdrifts. In the High Arctic, seedlings are very rare, and some grasses, saxifrages and buttercups never produce mature seeds, multiplying exclusively by vegetative reproduction from shoots on subterranean runners. This method of reproduction, and also that in which bulbs are transformed into bulbils which detach themselves from the plants, enables the latter to reproduce quickly in very low temperatures.

Insects that feed on green plants do not range as far north as those dependent on lichens, plant remains or animal food. Under conditions of snow and ice springtails perhaps form a food base, as they do in deserts, and no doubt snow-fleas, which are to be found in black masses covering a square yard or more on Svalbard, supply some birds with food. With the thaw in spring or early summer numbers of mosquitoes appear and, with blackfly and other blood-sucking diptera, make the short Arctic summer a hell for man and beast, and are possibly responsible for a certain mortality among their mammalian victims. All these diptera have aquatic larvae, whose numbers are perhaps controlled to an appreciable degree by the Arctic char and other freshwater fish, which also take mature insects flying low over the water. The diptera play an essential role in Arctic ecology by providing food for the millions of breeding birds and particularly their young, whether they be passerines, small waders, duck or gulls. The vast majority of these are summer residents, and there is little food for them when they first arrive at their Arctic nesting grounds, though most of them take a wide range of foods, including berries and willow buds. Nevertheless, despite their loss of weight during their migratory voyages of hundreds or thousands of miles from their winter-quarters, most arrive in sufficiently good condition to enter without delay into the vital activities of establishing territories, courtship and mating. Every operation is completed by the most economical methods and in the shortest possible time. Golden plover, for example, can breed north of 70° because both parents assist in rearing the young ones, and because the latter, developing extremely rapidly, are independent of parental care at a very early age; while the adults, in order to complete their molt into winter plumage before the autumn migration, shed their feathers so rapidly that they are almost incapable of flight while doing so. Even resident species, such as the rock ptarmigan, succeed in molting three times in the course of three or four months—from winter into summer plumage in June, from summer into autumn plumage, and finally into winter plumage again in August and September.

But Arctic birds breed only when conditions are right. In unfavorable seasons only 1 or 2 per cent of a particular species of summer resident in a particular area may breed. Similarly, resident species, such as snowy owls or eider duck, may not attempt to breed if the spring thaw is late or food supplies are abnormally scarce. Annual breeding, irrespective of conditions, is the exception not the rule in the Arctic. Snowy owls breed almost as far north as there is land, and ravens to 75° N in Greenland. Although some owls and ravens winter in or near their breeding area, the majority retreat to the southern fringe of the Low Arctic, but nevertheless return to hilly regions clear of snow with or before the return of the sun. A few joints frozen off its black toes are apparently no handicap to that remarkable bird the raven, as much at home on desert sands or in the 20,000-foot fastnesses of the Himalayas, as on the ice-fields to which it seems so ill-adapted when playing jackal to polar bear, wolf or fox, picking over their droppings or flying out over the ice to the carcass of a seal killed by a bear. "*Nanu-qapa?*" shouts the Eskimo to such a raven—"Is there any bear?"

More generally resident in the High Arctic are the rock ptarmigan, inhabiting the biotope of the hares as far north as 75° in Greenland, and retreating only short distances south during the winter. Although enjoying the luxury of roosting through the polar night in snow-tunnels reported to be 65 degrees warmer than the night air, the ptarmigan must experience considerable difficulty in procuring sufficient food during the winter to maintain an adequate insulating layer of fat under their skins, since they lack the hares' prognathous teeth and strong feet. From May to June, throughout the summer, the ptarmigan can fatten up on bistort (snakeweed) bulbils rich in protein, despite the stresses of the breeding season and the succession of molts; but heavy falls of snow in October or November oblige them to change to their winter diet. Some plants, protruding through the thin covering of snow on windswept plateaus, are easy to obtain; but to secure those rich in fats, such as willow, mountain avens or purple saxifrage, they must scratch down through deep snow. And this entails a critical outlay of time during the short period of "twilight" available on a winter's day in the High Arctic. A ptarmigan feeding on avens, for example, takes only the leaves, and does so very precisely, snipping them off one by one near their bases. This technique may be suited to a summer's day of upwards of twenty-four hours, but the rate of consumption cannot be increased to suit winter conditions. During the winter, therefore, the buds and tips of willow are the ptarmigan's main source of food, for in a couple of hours it can fill its crop to distension—which it never does at other seasons—with some 10,000 fragments of willow. No doubt, during the latter half of the winter, when the snow in caribou country has become too compacted for the ptarmigan to scratch down to willow and berries, they associate with the deer and take advantage of the vegetation exposed by scraping hooves and antlers, just as the ptarmigan of the Soviet Arctic associate for this purpose with the reindeer, and the red grouse of the Scottish Highlands with the red deer.

Although the snow bunting is the small bird which one

immediately associates with all that is Arctic it, like the wheatear, is in fact a summer resident only, though nesting within six degrees of the Pole. Only one small passerine, Hornemann's redpoll, whose range includes Greenland as far north as 77° and possibly Ellesmere Island, has been able to adapt itself to permanent residence in the High Arctic. Very little is known about its habits, and I commend its study to some hardy naturalist. During the winter, apparently, small flocks of a dozen redpolls work around the 1,000-foot contour in the almost inaccessible highlands of the interior. There, despite the very low inland temperatures, they feed on seed-bearing plants protruding above the snow and on willow and other scrub on mountain slopes from which the winds sweep the bulk of the snow. In the spring, however, they migrate to the coast, when the thaw results in the formation of an ice-crust platform from which they can reach the taller vegetation.

The birds of the true tundra are so varied and numerous that a comprehensive account of them, and of such mammals as wolverines and barren-ground bears inhabiting the less Arctic regions of the tundra, must await a later volume in this series.

16: Arctic Foxes and Hares

There are two types of Arctic fox, little larger than domestic tom-cats, weighing no more than 25 pounds in the season of plenty, and only a quarter of that when near starved during the winter and early spring. In one the color of the winter fur is smoky blue, in the other white. In the main they inhabit different environments, or biotopes, and eat different foods; so that although they interbreed, producing cubs that are blue or white or in a minority of cases gray, they can perhaps be considered distinct races, though I suspect that too rigid a division has been drawn between their environments. However, the blue fox is the coastal fox, rarely visiting the tundras of the hinterland, and subsisting predominantly on the produce of the sea and the left-overs from polar bear kills during the winter, and on berries and on sea-birds and their eggs during the summer, when their largest populations are concentrated near the vast breeding places or loomeries of little auks on the cliffs. The white fox is the tundra fox, preying on lemmings in particular, but also on hares and ptarmigan and other birds. As the blue foxes play jackal to the polar bears, so the white foxes scavenge on the caribou and musk-ox kills of the polar wolves; though since the latter usually appear to be in a semi-starved condition, pickings for foxes must be lean, and perhaps no more than the frozen blood-trail of a wounded beast. But according to Eskimos, foxes are themselves capable of killing very young caribou calves temporarily left by their mothers.

Since the white fox could probably not survive in a tundra environment without the lemming—its main source of food—it is a reasonable hypothesis that it is a specialized form of the blue fox, which adapted to life on the tundras after the lemmings had colonized them at some time before or after the last glacial era. Certainly over immense areas of the Arctic the association between type of fox and type of prey is extraordinarily close. Where lemmings are abundant the fox population may consist almost entirely of the white type—99 per cent in Arctic Canada, 96 to 97 per cent on the Siberian tundras; while on the Pribilof and Komandorski Islands, where there are both lemmings and loomeries of little auks, the blue foxes, which predominate over the white, are reported not to prey on the lemmings, though this may be because the fur seal rookeries provide them with an ample supply of carcasses. However Nature, with her customary contrariness to our neatly packaged hypotheses, has also arranged that on the Svalbard islands, where there are no lemmings but immense loomeries of little auks, 99 per cent of the foxes are nevertheless white; while on Iceland, where there are again no lemmings but millions of cliff-birds and also large numbers of duck and geese, the few foxes may be either white or blue. So we are left with the one hard fact that in all parts of their range the tundra foxes are very rarely of the blue type.

In order to survive the hungry months of winter the blue foxes must be ever on the prowl along the beaches, scavenging the tide-line for sea-urchins, crabs, small crayfish and cast-up fish, or catching tysties in the tide-cracks or, and above all, be ever in attendance on bears. It is at least arguable that if there were no polar bears the population of blue foxes in the Arctic would be hard put to it to survive. Wherever the polar bear travels, with him go the foxes. Almost every bear, hunting along the tide-cracks in the fiords or on the ice-foot, is followed by one or more foxes, for whom the seals he abandons provide feasts. At worst the undigested matter in the bears' droppings may save them from starvation. In this essential association with bears it is the latter who do the work, for while the bear is conducting his prolonged and painstaking stalk of the seal, the fox curls up to sleep on a nearby hummock or berg. However, the association is not entirely free from danger for the fox, who may venture too close to a carcass while a bear is feeding on it and be killed by a swipe from his sledgehammer paw. Every stranded carcass of whale or walrus is the goal not only of bears but of dozens or scores of foxes—ghost-like in the winter half-light—and is surrounded by a maze of tracks and pitted with holes where the foxes have burrowed into it, for despite their relatively weak jaws they are capable of gnawing holes even in the exceedingly tough, 2½-inch thick hide of a walrus, in order to feed on its blubber and flesh.

Large numbers of foxes accompany bears out on to the pack-ice. When, for example, Nansen was some 250 miles north of the New Siberian Islands in the middle of December, bears accompanied by foxes suddenly appeared; while Peary came upon the fresh tracks of a fox near a lead when he was north of 87°, and therefore some 300 miles from the north coast of Greenland. In the early spring indeed foxes may be permanent inhabitants of the pack far off the land. Although they are good swimmers, hundreds are unable to return to land when the ice breaks up, and are carried on the floes a hundred miles or more out to sea. Stranded on the pack in a starving condition, they are bold enough to scramble up on to the decks of frozen-in ships in search of scraps and offal. Only foxes and polar bears reach Iceland with the drift-ice; lemmings, ermines, hares and wolves have never done so.

Although these "marine" foxes must be dependent upon polar bears for much of their food during the lean months, they are better able to fend for themselves when the ringed seal pups are born in the early spring, for they are no less skilled than bears in locating occupied aglos (breathing holes) and in digging down to them with such astonishing rapidity that the snow streams back from their scratching paws as if from the rotors of a snow-blower. Considerable numbers of ringed seal pups, and even an occasional adult, are known to be killed by foxes in the Canadian Arctic, though it would seem an impossible feat for such a diminutive animal to drag from an aglo even a small adult seal, weighing at least a hundredweight. The victims are skinned and cleaned of every vestige of blubber with a machine-like efficiency, which the fox achieves by biting the skin around the seal's mouth and then drawing the entire carcass, including the flippers, through this aperture, without making any further incisions in the skin.

During the summer months the blue foxes obtain their food at the loomeries, scrambling about on the almost sheer cliffs hundreds of feet above the breakers, in quest of the crevice-nesting auks and their eggs. In the late summer they lie in wait for the young guillemots which, when they fledge, "float" down to the sea from the nesting ledges high up on the cliffs. Some fall short of the water on to the rocks and these the foxes snap up if they are not forestalled by the glaucous gulls, which take as great a toll of these young auks in the Arctic as the great-blackbacked gulls and bonxies do in the Shetlands. The foxes also take the nestlings of Arctic terns and ivory gulls. These are not for immediate consumption only, for both blue and white foxes store food in rock crevices or under boulders, covering their caches with turf, moss, sand or gravel. Alwin Pedersen found one blue fox's cache in a rock crevice 18 inches below the snow, which contained 36 little auks whose heads had been bitten off, 2 young guillemots and 4 snow buntings, together with a large number of auks' eggs—a sufficient store of food for a month or more. The frozen bodies had been neatly arranged with tails all pointing the same way in a long row at the bottom of the crevice, while the eggs had been heaped against the face of the rock. W. H. Feilden (quoted by G. Nares) has described how although his party actually saw only two pairs of white foxes during an expedition to North Grinnell Land, the ground was honeycombed with the holes of caches, each of which housed the bodies of 20 or 30 lemmings, and in one instance 50, covered by a little earth. One cache contained the greater part of a hare. "The wings of young brent geese were also lying about; and as these birds were only just hatching, it showed that they must have been the result of successful forays of prior seasons, and that consequently the foxes occupy the same abodes from year to year."

Apparently some, or most, foxes do not open up their caches during the hungry months of winter, but during the spring season of famine before the lemmings begin breeding; and one three-legged Greenland fox is known to have traveled a hundred miles north to winter in Washington Land, and to have returned in April to open up its caches near Etah. That a fox should deliberately abstain from drawing upon its caches during the winter months seems very remarkable; but the habit of storing up reserves during the summer, when food is in excess of immediate requirements, would surely have long since fallen into disuse were it, as has been suggested, no more than a squirrel-like hoarding of trifles to be forgotten. Perhaps the habit originated in the practice of bringing food for the cubs to the earths, which are only in use during the summer, and may be used by generations of foxes. Early in July, for example, James Clark Ross opened up a burrow, comprising several passages from 6 to 12 feet long converging on a single chamber, in the sandy margin of a lake, and found that both the passages and the cell contained numbers of lemmings, several ermines, and great quantities of the bones of hares, ducks and fish.

Freshwater fish, when available, may form a useful addition to a fox's diet and, many years ago a Russian observer, Nosilov (quoted by S. I. Ognev) described the behavior of seven foxes on the shores of a Novaya Zemlya lake covered with new ice:

> The foxes were playing, scampering about on the ice, hurling themselves against each other, landing on their paws, leaping high in the air, again falling on the ice, rolling on the surface and romping about like puppies at play. One leaped high in the air, crashing through the ice near the shallow shore, immediately reappeared on it with a fish in its teeth. At first the others were apparently surprised by this, but the next moment all the foxes began to fish. They would watch an unwary loach, as a cat hunting a mouse, as it drifted too close to the shallow shore, lying in wait in a curled-up position like a heap of snow and eyeing the fish beneath the ice. Only at the proper moment did the heap of snow suddenly leap high above the surface to fall on to the ice and crash through it to catch the fish.

Since the habitat of the white foxes necessarily coincides with that of their main prey, the lemmings, they range as high as 3,000 feet on many of the Arctic islands, and very much higher on the Greenland ice-cap, where lemmings' nests have been found at altitudes of several thousand feet, and where hares inhabit the nunatak peaks protruding from the inland ice. The foxes perhaps traverse the ice-cap with some regularity, for they are quite frequent visitors to the American weather stations, no doubt following the tracks of men, dogs and motorized sleds up from the coast and from one station to another. Almost three and a half centuries ago the Dutchman, Van de Brugge, noted in his journal of a winter on Vesterspitzbergen that he and his six companions were much disturbed at night by the constant scratching of numbers of foxes on the roof of their hut, and also by bears fighting over the carcasses of other bears the men shot. And ever since those early days of Arctic exploration, man and his food supplies have invariably attracted all the foxes in the neighborhood to the huts of trappers and the tents of expeditions.

During the summer the white foxes have no difficulty in obtaining ample supplies of lemmings, which are often above ground at this season or in shallow burrows in the newly thawed ground, especially where frost has churned up bare blisters among the plant cover, enabling the lemmings to tunnel into the sides of the top-soil warmed by the sun. Lemmings also frequently cool off in puddles of snow-melt or in the spongy saturated sphagnum moss. The fact that they are mainly nocturnal does not afford them protection during the light Arctic nights, and a fox has in any case acute hearing. Even after an early autumnal fall of four or five inches of snow the fox can hear the lemmings tunneling through it swiftly and effortlessly, butting it with their blunt heads, scraping with fore-paws, throwing it back with hind paws. Zigzagging over the snow at a leisurely trot, the fox suddenly halts and cocks his head on one side, listening for their activities. Then, after a moment or two's alert attention, he makes three or four stiff-legged springs, leaps high in the air like a diver from a springboard, and lands with nose and paws together on the spot he has selected. Breaking through the snow's crust, he snaps up his victim and kills it with a sharp nip. But since the fox has already stored his caches, and the numbers of lemmings are still in excess of his requirements, three-quarters of those he catches may be dropped on the snow; and although he may contemplate them for a moment or two and then pick them up again and bury them in the snow, they are probably forgotten.

In the autumn, when the white fox is in need of an abundant supply of food to nourish the growth of a thick winter coat in November and December, the lemmings are again venturing above ground from their tunnels among the roots in order to secure grasses and sedges. These they chop up into 1 inch lengths for their winter nests, which are egg-shaped and about 10 inches long and lined with their own fur or with the hair and wool, when available, of caribou and musk-oxen. But as winter draws on, and the several feet of snow above the lemmings' nests becomes packed down and frozen, the foxes are hard-pressed for food. This scarcity of food on the tundra during the winter and early spring prohibits white foxes from being tied to a permanent home in the form of earth or hunting territory and they roam far and wide—as indicated by the recovery of one marked fox at a distance of 600 miles from its birthplace. Although they have relatively narrow paws, ill-adapted to deep valley snow, the foxes can accomplish long migrations over the sea-ice or compacted tundra snow. Many go down to the coast and compete with the blue foxes for food along the beaches and on the ice-fields and some, drifting with the pack-ice, invade distant coastal and island biotopes of the blue foxes. Some of these invaders, unable to return to their tundra homes when the ice breaks up in the spring, establish new colonies; though it is suggested that, because they do not possess the scavenging know-how of the experienced blue foxes, and therefore suffer a high mortality from starvation, such colonies can exist for a number of years only if regularly reinforced by fresh invaders.

Arctic (white) fox

Not only have the white tundra foxes to contend with the perennial scarcity of winter food, but also with the periodical catastrophes to which the lemming populations are subject, and those occasional climatic fluctuations which prove disastrous to all life on the tundras. Although lemmings have been able to colonize the Arctic tundras—wherever there is adequate winter snow cover—from Scandinavia and Siberia to northern Canada, the Arctic islands and northern Greenland, they have been able to do so only because the tundra climate is of a stable continental type: dry and cold during the winter, mainly dry and cool during the summer. These are therefore the conditions that favor the tundra foxes, whereas the blue foxes benefit from the current warming up of the Arctic and the resulting relatively mild winters and longer seasons of ice-free coastal waters. But from time to time, even in the High Arctic, unseasonable thaws or heavy rains during the late autumn or winter are followed by the inevitable freeze-up. This seals the tundra in a thick layer of ice, encasing the herbage in an impenetrable covering which neither lemming nor hare, caribou or musk-ox can break through, in order to feed on the willows, lichens, grasses and sedges. Such sudden climatic changes periodically decimate not only the various populations of rodents but also those of the foxes that prey on them.

The effect of these sudden thaws or rains on the lemmings is that the greater proportion of a local population are either drowned in their tunnels or are saturated and die of cold, or starve because they cannot feed when the snow cover is broken down and subsequently freezes into hard ice on the cold ground. If the numbers of lemmings are decimated in this way, then those foxes that have stayed to winter on the tundra are in such poor condition by the spring that they are unable to rear cubs, and may kill and eat them. However, the after-effects of these untimely winter rains are not wholly unfavorable, because they replenish the tundra's water reserves and regenerate its vegetation; while in the dry periods that follow, the surviving stocks of lemmings multiply and the numbers of foxes increase again. That life is a struggle for the tundra foxes more often than not is reflected in the fact that to compensate for the bad years they are more prolific in the good ones than the blue foxes; for the latter—whose populations vary only slightly from year to year—rarely produce more than ten young in a litter, whereas a white fox may produce as many as twenty.

It is not easy to determine to what extent the white foxes prey on hares. Although leverets are regularly taken during the summer months it is possible that adult hares, which have extraordinary powers of acceleration, are too agile for them. On the one hand, Freuchen—than whom no man was more intimately acquainted with life in the High Arctic—states that many hares are caught by both foxes and wolves. On the other hand Alwin Pedersen, and most other trappers and natives of the Arctic, are equally positive that hares are too fleet of foot to be killed in any numbers by either foxes or wolves.

So far as wolves are concerned this may well be the case, for one will lope across a hillside alive with hares on a midwinter day and ignore them. However, when Otto Sverdrup encountered a pack of ten or twelve wolves on the pressure-ridges in the Axel Heiberg Sound on a day early in May, he noticed that the ice-foot bore innumerable tracks of hares leading out to the drift-ice where, he surmised, they passed the day in retreat from wolves and foxes in holes among the hummocks. Sverdrup had some curious experiences with hares on the Canadian Arctic islands, and observed that when they were driven they would run for short distances erect on their hind legs. On one occasion he walked very slowly towards some scores of hares nibbling at the grass on a small hill, in order to discover how closely they would allow him to approach them, and he describes, in *New Land*, how when they caught sight of him they gradually collected together:

> Before long they were an unbroken white mass, with their heads inwards and their tails out. There were so many of them that there were several rings, one inside the other, and it was a life-and-death matter to be in the innermost ring—at least, so it appeared to me, for they made the greatest commotion about it. They pushed and fought and bit each other till they screamed aloud, all the time slowly revolving, something like a millstone.

"So this was the square of the arctic hares!" he commented—comparing it to the defence formation adopted by musk-oxen.

Taking into account their size—2 feet or more in length and from 7 to 15 pounds in weight—the polar hares may almost equal the lemming hordes in mass, for their numbers in one region of Svalbard were estimated at around 5,000 to the square mile. Though descending to lower levels near the sea during the summer months when, according to Eskimos, they travel long distances to "drink" from snow-drifts, they avoid marshy flats and are essentially hill fauna up to an altitude of 3,000 feet, especially in the winter when the winds sweep the high tundras free of snow. Although the herbage may be sparse at these higher levels, the hares thrive on it, possibly because the autumn frosts strike the plants suddenly when fully ripe, with the result that the bulk of their nutritive elements are preserved in frozen form beneath the snow. A quarter indeed of the tundra's plant leaves, sedge and cotton grass may remain green throughout the year in this way. The hares' basic foods are lichens and the leaf-buds, twigs, bark and roots of the creeping willow, though those hares inhabiting coastal tundras will gnaw at the tough leathery kelp and, if starving, at meat-bait in traps or blubber stored in caches and *iglooviuks*. Under winter conditions they are able to obtain supplies of willow by scraping away the snow with their prognathous teeth and blunt claws, which are much stronger than those of other races of hares. Moreover in hares of High Arctic regions—and in Greenland their tracks have been seen within 450 miles of the Pole—the four incisors are abnormally long and protrude obliquely. This development, coupled with jaws that act as long-nosed pliers, must be an additional aid to them in tweezering out willows and saxifrages from deep snow, nipping off buds and twigs from frozen ground, or extracting them from crevices between stones. Even when the snow crust is frozen they can scent out plants covered to a depth of one or two feet. They then break the crust by stamping with their forefeet, and remove the chunks of frozen snow with their teeth or push them aside with their noses.

No animals are better equipped to withstand the super-severe winters of the high tundras. Even during blizzards they rarely tunnel into the snow for shelter, but crouch in crevices in the rocks and allow themselves to be covered by the drift. So long as they can maintain body temperatures at an adequate level, by the intake of sufficient food, the High Arctic hares are impervious to cold or storm but, deprive them of their food supply, and catastrophe ensues. Periodic epidemics also decimate them in tens of thousands and provide that control over their teeming populations which predators apparently cannot do.

17: Lemmings—the Manna of the Tundra

Lemmings have been described as the manna of the tundra, and they are in some degree as basically essential to its animal life as the euphausians are to that of the sea. But unlike the euphausians, which are wholly beneficial to their environment, lemmings can be most damaging to theirs.

Two races of lemmings inhabit the High Arctic—the brown lemming which is, however, absent from Greenland, and the collared or varying lemming which molts into a winter-white coat. The two races may occupy the same locality, though in this event the collared tend to inhabit the ridges of high ground that penetrate the cotton-grass flats of the brown. But it must be made clear at the outset that though in his book, *The Lemming Year*, Walter Marsden lists some eighty papers and books concerned with various aspects of the lemmings' life-history, and though scores of zoologists and naturalists have studied and experimented with lemmings in many countries, their theories and observations, together with those of the native peoples, are so conflicting, and the habits of lemmings apparently so at variance from one region to another, that one hesitates to state categorically that any lemming performs this action or behaves in that way.

However, we can begin with an undisputed fact—lemmings do not hibernate. The only true hibernators in polar regions are the ground-squirrels of the Low Arctic, which over-winter in shallow chambers less than 3 feet long and only a few inches above the permafrost, though a colony's complex system of burrows, inhabited year after year, may extend for upwards of 70 feet. Where the soil thaws to considerable depths in the banks of lakes and rivers the squirrels may be able to tunnel 8 feet below the surface; but even shallow burrows are warmed by their snow cover, 5 or 6 feet thick. Since the layer of soil above the permafrost also freezes in September the squirrels are obliged to prepare for hibernation as early as August, and begin collecting nesting materials of grasses and lichens, together with upwards of 4 pounds of food per nest for use in the spring until they are able to venture above ground again in April.

The lemmings are also dependent upon a snow cover of at least 3 feet to heat their winter nests to temperatures variously estimated as 50 or 70 degrees F higher than those prevailing above ground. We have already noted the disastrous consequences of unseasonable thaws or rains. Similar disasters occur if the lemmings are unwise enough to prepare winter nests either in exposed sites from which the snow cover may be swept away by gales or in localities which, though providing good feeding in the summer, do not have an adequate snow cover in the winter. It would seem true to say that, among rodents, only hares could exist under Arctic conditions without cover, and that lemmings and other small rodents could not—nor therefore could the foxes and such raptors as snowy owls which prey on them.

During the summer the lemmings' main foods are, depending upon locality, grasses, the leaves of creeping willows and birches, and various berries and fungi. During the winter some of these are also available beneath the snow, together with roots, including the long tap-roots of liquorice (*Hedy-*

sarum) and such small bulbs as those of *Polygonum*; but the succulent skins of the younger twigs of creeping willow are preferred, and a nest is often situated near a mat of willow. To obtain these winter foods the lemmings engineer extensive tunneling systems which, as local supplies become exhausted, may reach 60 feet in length with many blind lateral branches.

Lemming

Reports, however, that they may extend for hundreds of yards, and that they are driven through the snow-cover across deep rivers, require confirmation. The tunnels, which are revealed only when the snow-drifts disintegrate after midsummer, radiate from the nest. One tunnel will lead to a midden, in which the droppings may form a pile 8 inches high and 15 inches across; another to a larder stored with the peeled white wands of young willow twigs and old "branches" with bark untouched. Since Arctic lemmings are only from 5 to 6½ inches long, and weigh only from 2 to 4 ounces, and can moreover contract their bodies to turn in very much less than their own length, the tunnels are mere pipe-holes. Nevertheless ermines can insinuate their sinuous forms into them and are the lemmings' most persistent predators. Almost certainly ermines could not survive in the Arctic if there were no lemmings, for though widely distributed, they are nowhere numerous, nor are they well equipped to withstand low temperatures, quickly succumbing when trapped above ground.

It is presumably as an aid to tunneling through snow that the collared lemming develops in the autumn a thick horny pad (detectable in the young ones when they are only ten days old) on the underside of the two powerful middle claws of its forefoot. This pad, which reaches a length of 2½ inches and protrudes beyond the true claws, is apparently eventually worn off by use, and is not, as formerly believed, shed when the lemming molts out of its winter coat. But if it is in fact an auxiliary tunneling tool why are the brown lemmings not equipped with it? The answer may be that though the latter actually burrow more extensively than the collared lemmings because the snow lies to a greater depth in their valleys and flats, the collared require longer claws with which to excavate the densely compacted snow of the windswept uplands.

Although there is evidence that some lemmings of both northern races actually begin breeding during the winter, the first litter is usually produced beneath the snow in March. This is always the most significant event of the tundra year, for this first litter of half-a-dozen young is succeeded, after a lactation of only 14 days and a gestation of 21 days, by others at regular intervals until August or September, providing an inexhaustible reservoir of food for predators and raptors. However, periodically, usually at three or four year intervals, the produce of these early March and April litters are themselves able to reproduce the same summer—when only half an ounce in weight. Various theories have been advanced to account for these more or less regular cycles of abnormal sexual precocity, and these will be considered at the end of this chapter; but at this stage we shall treat them as basically the consequences of a bumper food supply, following a favorable series of early springs and late autumns in certain areas of the tundra. However, the affected areas are apparently localized with, for example, only one of two separate populations of lemmings on either side of a high ridge involved—though I find this difficult to credit.

Nevertheless, the explosion of numerous local populations has catastrophic repercussions not only on the general ecology of the tundra, but on that of regions outside the Arctic, in subsequent years; and a great part of the economy of tundra life may be said to revolve around this phenomenal sexual urge of the lemmings. In the winter following their population explosion an enormous number of lemmings are concentrated in the tunnels under the snow, and by March they are already searching for food on top of it, having exhausted all the supplies of roots and vegetation beneath it. Their abnormally early appearance above ground enables the vixen foxes to fatten into prime condition to produce large litters of cubs. The larger the litter, the greater the proportion of cubs surviving to maturity. End-result—the population of tundra foxes increases; and perhaps not only that of foxes, for in these peak years wolves are reported to feed exclusively on lemmings, ignoring their normal quarry, the caribou. Even the caribou themselves have been observed chasing and eating lemmings! The effects of the lemmings' explosion extends indeed to the most unexpected quarters. In Alaska, for example, the lemmings are feeding on the sedge beneath the snow during the winter and spring; but red-necked phalaropes rely upon a lush growth of drooping sedge to conceal their nests, and if the sedge has been abnormally "grazed" down by an extra large winter population of lemmings, many phalaropes are unable to nest.

The March swarming of the lemmings also attracts a variety of predatory birds, which are normally desperately short of food in the early spring before the arrival from the south of the millions of summer resident birds. In ordinary years skuas and snowy owls gorge on lemmings when the melting snow exposes their tunnels towards the end of April. The later the melt the more difficult it is for raptors to obtain sufficient lemmings to bring them into condition to breed; but an early swarming results in larger clutches of eggs from ravens, glaucous gulls, pomarine and long-tailed skuas, and especially snowy owls, whose normal clutch of four eggs may be more than doubled. But here, a word of caution. In keeping with the lemmings' reputation for providing the most awkward contradictions it must be emphasized that the flocking of raptors to that area of tundra in which a population explosion is imminent often *precedes* the lemmings' actual swarming above ground! In a year of plenty both ravens and skuas may feed exclusively on lemmings; the pomarine skuas not only swooping down on them from a height of 500 feet, but also hunting them on the ground, tearing up their runways with their beaks. But at all seasons of the year the snowy owl is the raptor in chief, and is therefore the most affected by the fluctuations in their numbers. In years when lemmings are scarce, owls may not attempt to breed, but when they are numerous, as many as eighty-four owls on their nest-mounds have been counted from a plane, each mound on its frost polygon, along a 100-mile stretch of the Alaskan coastline towards

Barrow. As in the dry valleys of Antarctica, so on the tundras, the contraction of excessively chilled soil results in the formation of polygonal plates 10 or 12 feet in diameter, surrounded by wedges of ice several feet deep, which melt into angular ditches in the summer. Built up over a period of years from accumulations of droppings and refuse, the owls' mounds stand 3 or 4 feet high and, when unoccupied, may be used by foxes as look-out stations for lemmings. Conversely, Stefansson has described how, when several foxes were digging up lemmings under the winter snow on Banks Island, here and there owls would be watching their activities from their mounds. "When a fox had traced a lemming and was about to plunge for it through the snow cover, an owl would silently approach, and the moment the fox had caught the lemming the owl would violently attack the fox with the obvious intention of scaring it into dropping its prey."

Throughout the second summer, and especially in July and August, the lemmings continue to proliferate, and the catastrophe nears its climax. By mid-August, according to some accounts, there is scarcely a square yard of tundra in an affected area that is not pitted with their burrows and runways. Large parts are literally eaten away and the vegetational complex of valley bottoms and the lower slopes of hills is considerably altered. Naturally therefore, in such conditions, the local population of lemmings begin to emigrate, not en masse initially, but in twos here and threes there; and although the total numbers involved may ultimately be huge, they do not travel as one army, but ramble aimlessly in divergent directions, with no apparent goal. It would be eminently reasonable to attribute their emigration to the conditions of an abnormally large population eating itself out of food; but the fact is that many, or perhaps

Snowy owl about to devour the lemming he has captured

most of these dispersals begin *before* the vegetation has been eaten out—though this does not rule out the possibility that supplies of certain favorite plants, essential in their dietary, may have been exhausted.

An alternative reason for these dispersals may lie in the

peculiar nature of lemmings. Superficially, they appear to live in large colonies, but their burrows are in the main unconnected, and a colony is in fact an aggregation of a large number of independent individuals, who have no collective warning system and who fight bitterly among themselves. During the mating season the males fight savagely, killing and partly eating rivals; and it is perhaps significant that if a number of lemmings are caged together, only one survives. The remainder are killed one by one and part eaten. It is possible therefore that an individual intolerance of overcrowding is as instrumental as hunger in triggering off a dispersal. Lemmings may be ill-adapted to cope with any form of stress, for they apparently lack the necessary reserve of fat in the adrenal gland to manufacture an emergency production of hormones to counter stress.

Whatever may be the actual factor responsible for a dispersal, the lemmings virtually cease reproduction and move out, and within a few months of their exodus those foxes that do not follow them are starving, while the owls and other raptors are unable to breed in their traditional areas. Those lemmings inhabiting low flat tundra disperse to all points of the compass, and ultimately out on to the bay ice, and thence on to the pack-ice. If confronted by stretches of open water they attempt to leap from floe to floe, but the majority are drowned or fall victims to ravens, owls or foxes, which are reported to follow them along the beaches of Siberia and drift with them to within 5 degrees of the Pole. Those from the high tundras move down into the valleys, radiating in opposite directions from a single plateau, or crossing the same river in opposite directions from either slope. If the water is flowing swiftly they venture in with the greatest reluctance, jumping in and out from the bank and returning to the shore again and again, but slow-flowing rivers they swim without hesitation, with head and shoulders well above the surface and hinderparts submerged, as they paddle furiously with their hind-feet and steer with their fore-feet.

Only when the lie of the land forces the numerous, independently migrating groups and individuals to converge into a single army, thousands strong, do the legendary hordes of lemming literature become fact. Probably only in such circumstances do their notorious stampedes occur, and these appear to be restricted to the lemmings of Scandinavia. Lemmings on migration are much less shy and much more restless than when at home, and the majority of the migrating males bear multiple lacerations attributable to fighting. When forced to mass in the confined space of a defile or peninsula, their close proximity, their frustration at being deflected by the lie of the land from the way they wished to go, their unfamiliarity with an unknown terrain, and the absence of familiar burrows, all no doubt accentuate their natural irascibility and excitability, and result in panic. At this stage they may almost be said to exhibit symptoms of insanity as, teeth chattering with rage or fear, they spit at and bounce up at an intruder; and in this condition they stampede blindly in all directions at the slightest incident—uphill or downhill, across small streams or large rivers and lakes, where salmon, trout, grayling or pike snap them up; or even out to sea where, incidentally, they continue to be as aggressive to each other as when on land. Some Low Arctic lemmings cross the Arctic line, invading forests far south of the tundra and entering villages and towns. Some swarms cause extensive damage to crops, others pass on through regions of abundant vegetation without halting. In human terms their apparently goal-less migrations, suicidal in the context of the colossal mortality they involve, can only be interpreted as attempts to escape from the mass of their fellows and find that solitude, the craving for which may, as we have seen, have provided the original stimulus of their exodus from their overcrowded tundra colonies. Biologically, they must be regarded as a self-regulating mechanism to control their various populations, which in most rodents are controled by periodical epidemics. Despite the apparently prohibitive mortality entailed in these migrations they must include a survival value for, if they did not, the practice would have been eliminated through the medium of natural selection; and one must assume that a minority of the emigrants always survive to found local colonies in new country.

It has been fashionable to associate the population explosions of lemmings and other rodents with four and twelve year sunspot cycles; but those of lemmings do not invariably correspond with the latter. There were, for example, no large-scale migrations in Swedish Lapland between 1942 and 1959; though this is not to say that there were no movements in the mountains and in the vast unwatched solitudes of the tundra. Moreover, favorable series of springs and autumns, which we suspect account for the build-up in population, do not necessarily coincide with sunspot cycles. Again, while varying hares and rock ptarmigan are subject to similar explosions, these occur in ten-year cycles. The varying hare, for example, produces litters of three or four leverets up to five times a year. Although only a small proportion of these may reach maturity, the overall increase during the cycle may finally result in a population of 4,000 hares to one square mile of forest. Then comes the crash, either from famine in an over-grazed region, or from stress due to overcrowding, and the resulting hormone disorders to which the hares quickly succumb.

Another ingenious theory is that periodic deficiencies or excesses of certain trace elements in the lemmings' diet disturb their endocrine balance. Tundra mosses and lichens are, for example, reported to be enriched at twelve-year intervals with vitamins which stimulate hormone secretions in the lemmings' anterior pituitary glands. Over a series of years the build-up of these secretions increases the lemmings' virility and fertility to a point where the population explodes, and the fact that explosions are localized, with flanking colonies unaffected, is explained by local variations in the timing of the vitamin enrichment of the vegetation. Moreover, since the vegetation outside the lemmings' habitat rarely contains these trace elements, new colonies founded during their migrations die out, because, once "fixed" on these vitamins, the migrants cannot survive without them.

However that may be, it cannot for the present be demonstrated that the lemmings' activities are associated with sunspots, solar energy or radiation, or with any lunar or other occult cycles. On the other hand, recent research on lemmings in the Point Barrow region would appear to indicate that their blood contains an anti-freeze substance. This, while it would assist them to remain active throughout the Arctic winter instead of hibernating, apparently has adverse effects periodically, possibly when stimulated by unseasonably high temperatures, and attacks their central nervous systems, resulting in insanity or death.

18: Wolves, Caribou and Musk-oxen of the High Arctic

If a polar wolf were capable of regret he might curse the day, perhaps 20,000 years ago, that his ancestors followed the herds of caribou and musk-oxen as they retreated north with the ice-cap over the barren-lands and the sea-ice bridging the narrow straits and settled in the frozen islands of the High Arctic, especially West Baffin Land and Ellesmere Land—the main summer pastures of the caribou—and then crossed the ice of Smith Sound to the ultimate limits of this epic migration in northern Greenland, within 7° of the Pole. In the same way, today, the wolves of the Old World still regularly follow the herds of reindeer across the Siberian tundra to the shores of the Arctic Ocean, and out over perhaps 125 miles of sea-ice to such islands as Novaya Zemlya, Kolguev and Lyakov.

On the other hand, later generations of Arctic wolves may have considered that a sporting chance of surviving starvation during the cruel winters, far removed from the bulk of mankind, was preferable to the mercilessly persistent and barbarous persecution to which their kin have been subjected in the easier living conditions of forest and prairie. This persecution has reduced their numbers in the U.S.A. to a handful, though many survive in parts of Canada, and in Scandinavia to a mere score or two; while in Siberia the Russians, for all their excellent work in the conservation of polar bears and walruses are apparently no less inhumane when the animal is wolf. But, you may say, wolves are no better than rats—destructful vermin and wanton killers that cannot be tolerated by men with flocks and herds. Possibly—but that is what man has condemned them to be, by depriving them of their traditional prey in the lands he has settled. That is not their status in their natural environment. There, they do not ravage the herds of deer and musk-oxen indiscriminately. When not actively hunting for food they live in harmony with the other animals in their biotope, moving freely through the herds of caribou which, recognizing their enemy to be full-fed by his posture and behavior, walk leisurely out of the way. For that matter, a herd may actually stand and watch the pursuit of one of their number which has lagged behind.

To be honest, there is the utmost confusion as to exactly how wolves hunt caribou and which elements of the herd they hunt. One observer states that they rarely attempt to follow a herd that has stampeded because caribou have such speed and endurance that they cannot be run down: another that the herd is in most danger when it stampedes because in the mêlée individual caribou cannot see the wolves. A third affirms that wolves take half the annual crop of caribou calves; but a fourth that they prey mainly on old cows and bulls which, with their great weight of antlers and fat at the beginning of the breeding season, cannot run as fast as the calves and yearlings, and therefore lag behind the herd. A fifth claims that an adult caribou can beat off and even kill a wolf with powerful blows from its fore-hooves, assisted by some defence from its immense antlers. However, we may safely say that in the main it is the laggards—the sick, the old and the very young upon which wolves prey, and in so doing maintain the vigor of the herds.

Selecting a herd, a solitary wolf chases it for a mile or two. If there is no weakling to drop behind, the wolf turns away and tests another herd. Sometimes many miles are run before a crippled caribou finally lags behind its companions. Swinging along at a swift gallop, the wolf slowly closes the gap, and with a final spurt he is alongside. When L. Kumlien was in Baffin Land about 1879 he observed that herds of caribou were continually beset by packs of wolves, which kept a vigilant watch for any that strayed out of the herd, immediately running them down; but they were seldom able to inflict many casualties in a herd which kept together, since the caribou formed a circle, with the weaker members in the center, and kept the wolves at bay. Today, wolves are rarely seen in packs; and even in the early years of this century, when they were very numerous on some of the Canadian Arctic islands, a pack of thirty or forty was exceptional. Ten or a dozen was the normal strength, and such a pack would be composed mainly of the two parents and their progeny of various ages, hunting a territory of perhaps 200 square miles. In Greenland, where they were probably never numerous and seldom bred, a single wolf may now be the only one of its kind in a territory almost as extensive as the Scottish Highlands; and even if four wolves are attracted to a trapper's refuse dump they come as individuals from different points of the compass. Even in caribou country numbers of wolves are reported to die of starvation during the winter when they are unable to run down the herds, better equipped to negotiate deep snow and ice. In regions without caribou, as is now the case in Greenland, they must face extraordinary problems of survival, for to maintain life they require a wide variety of small prey, and there is never any excess to cache. The few remaining in Greenland must eke out their existence, and in three weeks one may go the rounds of a 200-mile beat, even venturing up on to the inland ice, on the endless quest for food. In the summer, sitting birds and their eggs, lemmings, leverets, fox cubs dug out of their earths; in the winter, foxes in traps, the remains of fish and crustaceans cast-up along the tide-line, a sick musk-ox, and carrion (but urinating contemptuously over an ox's carcass poisoned by trappers).

In the springtime their vagrant's diet can be supplemented

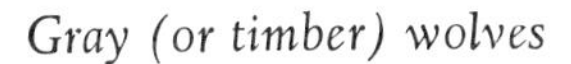

Gray (or timber) wolves

by seal pups, for according to Freuchen they are even more skilful than bears in scenting out aglos (breathing holes), though when hunting deer upwind a wolf perhaps makes more use of his ears than of his nose, for he will pause to stand with head on one side, one ear pricked and the other laid back, as if listening. Later in the season the wolves can hunt seals basking on the ice, and have indeed been reported stalking them so persistently on the West Ellesmere ice in some years that the seals would seldom lie out. Of man they are normally timid and shy, though bold enough to kill huskies in the camp-lines; and although there is some circumstantial evidence that a family of starving wolves may be driven to attack a man, I do not know of any indisputable first-hand account of such an attack, and Eskimos consider them harmless. However, Bjørn Saib, who made an unsuccessful attempt in 1964 to reach the North Pole on skis, was told by a Finnish-American member of the U.S. airbase at Alert, in the extreme north east of Ellesmere, that he only escaped from a pack of wolves on the airfield one night by leaping into his "halftrack," from which, according to his somewhat astounding claim, he shot twenty-seven of them.

To understand the distribution and problems of polar wolves we must examine the life history of the caribou which, as their main prey, have determined these. Wherever there are caribou, wolves are never very far away; and it has been roughly estimated that 20,000 square miles of tundra on a Canadian Arctic island may pasture as many or half that number of caribou, and form the hunting preserve of 150 or 200 wolves. If the caribou are wiped out in one area, the majority of the wolves move out to fresh hunting grounds. In the beginning the barren-ground caribou wintered among the muskegs and sparse forests of the sub-

Arctic zone on the southern fringe of the tundra. There they had to compete for food with the larger woodland caribou (now almost extinct) and had exhausted their reserves of fat and were very lean by the spring, when they set out on a migration of as much as 700 miles to their summer grazing grounds on the shores of the Arctic ocean, where they would arrive in time for the birth of their calves in May and June. At the end of the nineteenth century there were perhaps 2½ or 3 million caribou on the barren-lands. Today there are only some 300,000—the survivors of the indiscriminate burning of the lichens in their winter feeding grounds, for though they graze on fresh herbs and grasses during the spring and summer, lichens are their almost exclusive diet in the winter. The bi-annual migrations of *La Foule* (The Throng) were conducted on a grand scale. Their tens of thousands advanced on a fifty-mile front, and a single herd might require eleven days to make the crossing of a lake. Their myriad hooves flattened and bared the tops of the eskers—those geological curios whose steep-sided, 50 or 100 feet-high, "railroad" embankments of gleaming white sand and gravel snake across the tundra in all directions, and served the caribou for roads. As much as a hundred miles long, they must have influenced the caribou in their migratory routes. And the seasonal regularity of their migrations along traditional trails was a matter of life and death to the predators that depended upon their passing at the hungry season of the year—the gray wolves and foxes, the ravens and birds of prey, the native peoples of the barren-grounds whose entire economy and livelihood was geared to the coming of *La Foule*.

In the beginning, again, some of the herds that ventured across the ice, or swam the waters of the narrow straits, to the Arctic islands and northern Greenland did not make the return migration south in the autumn to the traditional wintering grounds among the muskegs and "little sticks." Possibly they were unable to recross to the Canadian mainland because of some climatic change. Possibly they were attracted to become permanently resident in the High Arctic by the abundance of lichens, especially the highly nutritious "reindeer-moss" *Cladonia rangiferina*, which forms a dense carpet over vast areas of the island tundras, and which is reputed to emit a strong fungoid scent, detectable by the caribou in two feet of snow. Despite the disadvantage of being long-legged and long-eared with angular frames, and thus, like wolves, exposing considerable areas of their bodies to the cold, the caribou were well-adapted to withstand the blizzards and low temperatures of the far north; for their short woolly under-fur is overlaid by a coat of long hairs which, being hollow tubes filled with air, act as insulators—and also buoy them up when swimming broad stretches of water. To support them on a surface of unstable snow-crust the two toes of their hooves are widely splayed, and also hollowed on the underside for traction over smooth ice. Through natural selection the caribou on some of the Arctic islands developed into smaller, shorter-legged races with almost pure white coats. The gray wolves that settled with them also tended towards whiteness, though perhaps few polar wolves are pure white and without some patches of hair with gray or black tips, while their litters of three, four or five cubs are born with gray hair curled like a husky's. Since this birth-coat is not weather-proof against moisture or drifting snow, the cubs spend their first six weeks in a hole dug in the slope of a hill or in a large fissure in the rocks high up in the mountains.

Although the lichen growth on the tundras is so abundant and widespread it is also extremely precarious. The cart-tracks imprinted on the tundra of Melville Island by Edward Parry's expedition in 1820 were visible to Leopold M'Clintock thirty-three years later, for hardly a speck of lichen or moss had grown in them. If over-grazed or over-trampled a sward of lichen requires between thirty and forty years to regenerate into caribou pasture from ½ to 1¾ inches long. If burned, regeneration may take a hundred years or even considerably longer than this. Thus in addition to climatic hazards, and to suffering what one would have supposed to be the crippling handicap of dissipating their entire reserves of fat during the rut in the late autumn, the various races of island caribou had also to contend with the problem of over-grazing. Two races in the Canadian Arctic islands whose stocks had gradually degenerated in isolation on poor lichen pastures—for they apparently ceased to be migratory—did in fact become extinct after a series of exceptionally wet winters during the last three decades of the nineteenth century. Wolves perhaps killed the last of them. In Greenland the surviving stock of caribou in the north-west succumbed to the effects of a heavy rainfall followed by a freeze-up in February 1901, as did the north-eastern herds soon after. The wiping out of the caribou deprived the wolves of their principal source of food, and they too became virtually extinct in northern Greenland, though a few lingered on until 1939. A crash in the hare population during the 1930s may possibly have contributed to their decline.

The wolves' other large quarry in the High Arctic are the musk-oxen, who formerly inhabited the rolling country at the southern edge of the tundra, typical of which were the rocky ridges, rounded and conical hills, small lakes and "little sticks" of pines and junipers, 3 to 8 feet high, scattered in small stands half an acre in extent and fifteen or twenty miles apart to the north of the Great Slave Lake. Some were still among the "little sticks" in the eighteenth century, browsing on the tips of the young pines in the winter or grazing on the edges of the frozen lakes where, west of Hudson Bay, Samuel Hearne frequently encountered many herds, some eighty or a hundred strong, in the course of a day; while as late as 1899 bulls could be found in riverine thickets of the barren-lands. But today, after decades of slaughter for their meat and hides, with countless herds decimated to the last ox by trappers, *voyageurs*, Indians, Eskimos and explorers—Peary alone shot more than 800 during his seven years in the Arctic—90 per cent of the world population of musk-oxen live on the Canadian Arctic islands and in north-east Greenland from Peary Land, within 500 miles of the North Pole, southwards to Scoresby Sound. These musk-oxen of the High Arctic are, like the caribou, slightly smaller than the mainland race and further distinguished by white foreheads, though in the bulls this feature is concealed by the horn bosses, which fuse to form a shield 4 inches thick.

The musk-oxen do not compete with the caribou for the latter's vital winter feed of lichen, for the oxen's main foods are the dwarf willows and birches, grasses and sedges which, since they do not possess upper incisors, they gather with their strong rubbery lips and tear loose with their sharp bottom teeth and tough upper palates. Though they usually migrate to summer pastures that have been rested under heavy snow during the winter, extensive snow-free stretches of grass may be grazed throughout the year; and if herds of caribou are also present on these summer pastures, they do not intermingle with the oxen. Where water is available from snow-melt, permafrost, glaciers or ice-cap, the oxen's primary feed of willows and grasses does not require additional moisture in the form of rain, though the oxen avoid regions where, because of lack of water, the birch withers too early in the summer. Throughout the summer they graze almost continuously night and day, pausing only to cud when their stomachs are full, or to move to patches of snow during the hottest part of the day in order to cool off; for although they get rid of an enormous mass of under-fur in the course of the summer, scraping it off against rocks and shrubs, they suffer considerably from excessive heat, not being able to perspire. Only in the extreme north of their range and in the mountains farther south do early autumn frosts preserve the nutritive elements of the vegetation by freezing. Elsewhere, it withers before the frosts, making poor feed for the oxen, who may only survive the winter on the reserves of fat they have built up by their industry during the summer. Their ability to survive is helped by their sedentary habits during the winter; but they are very lean by the spring and continue so until the burgeoning of the new herbage late in May.

Nevertheless, they can withstand the extreme Arctic conditions of Ellesmere Land and Peary Land. They have indeed been known to travel over Greenland's inland ice in search of fresh pastures, and bulls that have failed to retain their leadership over the herd are sometimes found on the nunataks, alone or in two and threes. They are admirably protected against the lowest temperatures and most severe blizzards; and are protected too—except for the edges of their eye-lids and the upper halves of their ears—against the attacks of black-flies and mosquitoes which drive the caribou to distraction in spring and summer, though mercifully the mosquitoes become sluggish when the temperature falls below 50 degrees F and do not fly when it is below 45 degrees F. Their dark-brown fleeces of guard-hairs, which are thicker at the tips than at the roots and upwards of 20 inches long, are underprooofed by a thick mat of wool fiber as fine in texture as a Kashmir *pashmina*, though heavily interspersed with coarse bristles. Sweeping the ground below their short legs, the fleece envelops them in the equivalent of an almost airtight fringed rug. Virtually every inch of their bodies, from ears and short tail, both hidden in the fur, to the soles of hairy feet, is insulated. Moreover, since a new growth of under-wool, working its way up through the outer hair, which is not shed, immediately replaces the old wool, they are never without complete protection.

However the yearling calves, who retain short woolly coats until their second autumn, do require protection from the elements. When a winter blizzard rages the oxen gather together in a herd, twenty or fifty strong, in a windswept place where the snow drives in a constant direction. There they form a circle, heads to wind, with the high humped shoulders of the adult bulls affording additional shelter to the yearlings huddled in the center behind the wall of fur. When the storm is at its height they close up still further into a compact wedge-shaped mass whose windward apex is formed by two bulls. And there they stand, without lying down, for days together, as the particles of snow become coarser and the depth of the current of flying snow increases until, in Peary's words, the "drift becomes a roaring, hissing, suffocating Niagara of snow, rising hundreds of feet into the air," and only the oxen's humped shoulders can be seen above the wreaths of snow. Warmed air is retained beneath the cloud of exhaled and transpired vapor; and this, when the storm moderates and the temperature is —30 or —40 degrees F, betrays the presence of a "yard" of musk-oxen or caribou at a distance of 5 or 6 miles.

Although no animal is better protected against the severest Arctic conditions, the musk-oxen have no defence against unseasonably high temperatures. Twenty-four hours' exposure to melting snow or rain saturates their under-fur, and if the temperature then falls again below freezing, ice forms among the guard-hairs. Wolves and foxes can break the icicles forming on their coats with their teeth and lick their fur dry; but the musk-oxen can only shake themselves, which does not remove the mass of ice entangled in their long hair. This may eventually become so heavy that they are unable to move, and are at the mercy of wolves. No less a danger to them than to the other herbivores is the thaw that may follow a heavy fall of snow, the subsequent freeze-up resulting in the formation of an ice-crust, which renders it difficult or impossible for them to dig down to the vegetation with the curved toes of their hooves and their hooked horns. In these conditions pregnant cows may be so undernourished by the spring as to be unable to calve, while the calves of many others may be born prematurely or die from lack of milk. Many of the younger beasts starve to death, as was the case in 1954 when almost the whole stock of one and two year olds is reported to have perished in the aftermath of a level fall of from 6 to 9 feet of snow over north-east Greenland. Such climatic hazards explain why musk-oxen have never colonized the more southerly regions of Greenland, where such unseasonable thaws occur with some frequency.

Nevertheless, it is somewhat surprising that musk-oxen thrive as well as they do in the High Arctic, for their rate of increase is very slow. Many cows breed only in alternate years, and the proportion of calves surviving the first few hours after birth, late in April or early in May, must be well below half and perhaps no more than one-third in late springs; for at birth the calf is only 20 inches long, weighs no more than 16 to 25 pounds and stands only 18 inches at the shoulder. The weather is still very cold at this time; nights are still longer than days; and there is no shelter from the piercing winds among the low rounded hills, stony and tree-less. Tottering on weak legs and shivering beneath its mother, it may be frozen to death during the first few critical hours, despite the partial protection of the cow's long fleece. If it survives this initiation into the polar world, then by the second or third day it is strong enough to keep up with the herd, and is nibbling at grass and herbs within a few weeks. The secret of its continued survival lies in the fact that the cow suckles it until a few weeks after the birth of her next calf, and for eighteen months if she misses a year.

Although a few wolves frequent musk-oxen country at all seasons of the year, an exhaustive search of Arctic literature does not suggest that the oxen have ever served as a major prey comparable to the caribou, despite one assertion that most adult wolves bear the scars of large wounds attributed to the oxen's sharp horns. That they often follow the herds, with a view to cutting out any old or sick beasts, or any calf temporarily overlooked by its mother, is not disputed, and Knud Rasmussen, the American explorer and anthropologist, records an instance of a young ox which, while being hunted by a wolf, became trapped by the legs between two large stones and was ripped open, with a single wrench of the wolf's jaws, from its thick gristly throat down through the chest to its diaphragm. Theoretically, a polar wolf should be capable of pulling down even a musk-ox bull, for he is a formidable predator, so large and with such a massive head and jaws as to be mistaken from time to time for a young polar bear. A male may tape 6 feet in length (including 18 inches of tail), stand 6 inches higher at the shoulder than a large husky, and weigh 150 pounds; and although they cannot injure a bear, a pack of three hungry wolves, snapping at his flanks, can pester him into abandoning his kill and lumbering away. But the fact is that the musk-oxen have evolved an efficient and virtually impregnable form of defence against wolves or perhaps, long ago, when they still inhabited the plains and forests, against some powerful carnivorous cat. It is doubtful whether they even fear wolves. Otto Sverdrup, for example, describes how when a pack of eight wolves attacked his dogs in camp, he shot one and wounded the other seven. The latter then retreated and unexpectedly encountered four musk-ox cows resting with their calves; but the cows did not even get to their feet, and though the wolves eventually formed a ring round them, they did not venture closer than 200 or 300 yards, continuing to howl from this distance for several hours.

However, if they are attacked by a pack of wolves or Eskimo huskies the oxen "form square"—actually an arrowhead or circular formation, the *karre*, developed perhaps from their storm "yard," with the bulls packed so closely as to be treading on each other's hooves, grouped round the cows and younger beasts, and the calves sheltering beneath their mothers' fringed bellies. With heads lowered, the protective screen of bulls presents a solid wall of horn and boss-plates against their attackers. In *New Land* Sverdrup has described the procedure when a pack of huskies attempts to break up the karre, though one suspects that he and other observers have somewhat exaggerated the precision of the operation:

> They stood at regular intervals one from another, with their hindquarters together, and their heads outwards. Then in turn, and with lightning speed, each one made an advance in the shape of a circular movement from left to right. At the same moment that an ox regained his place, his neighbor on the right passed out on a similar attack, and thus they went on uninterruptedly with almost military precision. As long as the maneuver continued, one of the oxen was always out on a movement of attack, endeavoring to spit or rip up one or more of his adversaries.
>
> The size of the attacking circle seems always to be determined by the distance of the enemy and the nature of the ground. As a rule, the animals advance ten or twelve yards from the square, and once I saw them make attacks to a distance of a hundred yards. The remaining oxen always cover the gap in the square, but immediately make room for their comrade when he returns from his round. Now and then, when the fight is a long one, they stop to breathe, and then begin again with renewed vigor. The greatest degree of precision is attained by oxen of the same age. I have seen herds of as many as thirty animals form a square, with the calves and heifers in the middle, and the bulls and cows standing in line of defence at distances as equal as the points on the face of a compass. When the defence forces of the line were no longer available, the reserve was mobilized; right down to two-year-old heifers. In such circumstances, of course, the movements were not carried out so regularly; and the discipline was less absolute. I noticed that sometimes the regular old fighters of the herd formed themselves into a kind of outpost, at twenty or twenty-five yards distance from the square. It sometimes happened that the whole herd first formed in a square, and that then one or two fighting giants would walk out to the out-

Musk-oxen in their defensive phalanx position

post's line; but, as a rule, their order of attack was evidently planned from the first. When once the animals had formed into a square, they remained at their post until the attack was repulsed, or the entire square had fallen. I have myself seen the last-standing ox make his sortie and then return to his fallen comrades. In cases where the oxen had to defend themselves against a single enemy they would sometimes form up in a long fighting-line, without cover on their flanks, and then stand forehead to forehead, and horns to horns.

It is conceivable that a sizeable pack of wolves, attacking the karre at numerous points, might be able to kill all the members of a small herd of oxen; and the American explorer, Donald B. MacMillan deduced from tracks on Ellesmere Island that one of a herd of bulls had been killed and eaten by a pair of wolves, after the smaller had attacked it head-on while the larger dragged the bull down by its hind-quarters. But the bulls are so nimble in their sorties and so expert with their sharp recurved horns—ripping up huskies and hooking them over their shoulders, to be horned and trampled to death by the cows and younger beasts—that such a killing must be extremely rare when a large herd of oxen has taken up its defensive position. However, it is possible that a bull might be pulled down and killed when making a sortie to some distance from the herd; and in this case the wolves would then retreat, and wait for the remander of the karre to break formation and stray away to graze, before returning to the carcass.

Alas, the karre, so admirably efficient against natural predators, proved to be an extravagantly suicidal defence against man the hunter. Over the centuries immense numbers of musk-oxen have been slaughtered, necessarily by Eskimos with the aid of dogs. And only after changing climatic conditions compeled the Eskimos of north-east Greenland to emigrate, were the oxen able to make that region their most populous. Although a musk-ox can out-gallop dogs for several miles initially it then settles to a steady and curiously choppy lope, with its long hair rising and falling with the motion of meadow hay billowing in the breeze. At this pace they can be overhauled by the dogs and brought to bay until the hunters arrive. There is no certainty as to how they will react on seeing or winding man. The bulls may become worked up, snorting and rubbing their horns against the inner sides of their legs; thereby impregnating the long hairs, which cover most of their muzzles from horn-bosses to nose, with a fatty secretion from a small gland just below the eyes. This, it has been suggested (plausibly if doubtfully) prevents the hair from blowing over their eyes while fighting. However that may be, they may then charge; and Baumann, a member of Sverdrup's party in King Oscar Land, described to Sverdrup how he was charged at full gallop by a herd of thirty oxen when they winded him at a distance of 100 yards—"So close on each other were their horns that they seemed to form a single, unbroken white line. The animals sunk their heads till they almost touched the ground, the steam stood out from their nostrils, and they snorted, blew and puffed." Nevertheless, on Baumann yelling and running towards them, the herd opened its ranks and allowed him to pass through; and repeated this a second time.

Usually, however, when approached by men and dogs, the oxen take up their traditional defensive formation and form square. Traditional but, as already noted, fatal in these circumstances because if one, and particularly the herd leader, is shot, the remainder stand around it and allow themselves to be slaughtered one by one. Indeed, if the hunter wants meat he is obliged to shoot the entire herd before he can obtain one beast. In *Nimrod of the North*, for example, Schwatka described how:

> As their numbers fall one by one the musk-oxen resolutely persist in their curious and singular mode of defence, presenting their ugly-looking horns towards as many points of the compass as their remaining numbers will allow. When only two are left, these, with rumps together, and, facing from each other, will continue the unequal battle against the enemy, and even the last "forlorn hope" will back up against the largest pile of his dead and dying comrades, or against a large rock or snow-bank, and defy his pursuers, dogs and hunters, until his death. While the little calves are too young and feeble to take their places in the front ranks (that is, until they are about eight or nine months old) they occupy the hollow square or interior space formed by their defensive

parents; but when their elders have perished in their defence, with an instinct born of their species, they will form in the same circular order and show fight.

It is a curious thing that if a single musk-ox is encountered and bayed, he will never remain satisfied until he has backed up against something, however small, to protect him from a rear attack. They have almost as much confidence in this trick as the ostrich has in hiding its head in the sand, for a rock no larger than a man's head will suffice if nothing better is conveniently near.

19: Problems of Whiteness

The higher the latitude the greater the number of mammals and birds with white fur or feathers and the paler-colored the race of a particular species, is a generalization applicable to the Arctic but not to the Antarctic. True, the snow petrel, one of the two petrels ranging nearest to the South Pole, is pure white; but the other, the Antarctic petrel, is brown and white; while the only other pure white bird in the Antarctic, the sheathbill, ranges no further south than Graham Land in the same latitude as southern Greenland. Nor is there any evidence that the off-white plumage phase of some adult giant petrels is more prevalent in the most southerly breeding colonies. In the High Arctic, on the other hand, snowy owls and ptarmigan are permanently or seasonally white, and Hornemann's redpoll is whiter than the mealy redpoll of the Low Arctic. Of the four color phases of gyr-falcons the whitest breeds furthest north in High Arctic Greenland.

Only in the High Arctic do wolves, lemmings and caribou assume white or near-white winter coats, while the High Arctic hare is permanently whiter than any other mammal, and the only one whose whiteness is comparable to that of snow, against which it is almost invisible. In the wan light of a winter's day, the hare's fur, silky and as soft as a powder-puff, so resembles snow both in color and texture that one may at first distinguish only the jet-black ear tips, apparently suspended in space above a hummock of drift-snow. The sealing Eskimo, believing that his quarry can detect white bear and fox skins, covers his sail with hare skins. If it is a matter of cause and effect, then either whiteness must have some camouflagic value for some prey and predators, but not for others; or it must be due to some climatic factor affecting some birds and mammals but not others. Supposing that polar regions, instead of being almost uniformly white in winter, were scarlet. Would bears, foxes and hares, owls and ptarmagin be white? I think not. Whiteness therefore has a camouflagic value.

Unfortunately for this proposition there is little evidence that this is the case. How often has one seen it stated that the pups of Weddell seals are rust-gray because, since there are no terrestrial predators such as polar bears or wolves in the Antarctic, there is no necessity for them to be camouflaged by the white coats with which Arctic seal pups are born. But in fact the creamy-white ringed seal pups are hidden from bears in chambers under the snow, and by the time they emerge on to the ice are light-gray "silver-jars"; the bearded seal pups out on the pack-ice are gray-brown; the hooded are a silvery blue-gray; and though the harp pups are white for the first three or four weeks after birth, their rookeries are not very often visited by bears.

When the collared lemmings are white they, like the ringed seal pups, are hidden in tunnels under the snow, in which they are heard or scented by foxes, not seen; though it could be argued, somewhat dubiously, that they "surface" more frequently from their lesser depth of snow than brown lemmings do, and that therefore their whiteness is of protective value against predators. Does this argument also apply to the brown lemmings in the High Arctic environment of the New Siberian Islands—the only members of their race to turn white in winter?

How does the tundra fox benefit by being white, for during the winter it is either a lemming eater or a scavenger, and has no significant enemies? And how do hares benefit by being white? During the winter their feeding grounds are predominantly stony flats and hillsides swept almost clear of the snow by the winds; while in summer their whiteness renders them extremely conspicuous against snow-free mountain slopes. In any case lemmings are the number one prey of most foxes and raptors, with hares a standby in lean lemming years; while the leverets, whose gray fur ought to camouflage them against a background of lichens and rocks, are much more heavily preyed upon than the adults.

Consider also the High Arctic ptarmigan. Its white winter plumage begins to develop in August, but has no protective value because snow rarely covers the ground at this season; nor does it by the middle of September, when the ptarmigan are pure white. Indeed, as soon as the white feathers predominate early in September, the ptarmigan become extremely conspicuous on snow-free ground. Dubiously again, it might be argued that for a limited period in early autumn the retention of the dark feathers on the back renders them inconspicuous to such birds of prey as snowy owls and gyrfalcons, searching for them from on high. It has also been suggested, in all seriousness, that ptarmigan actually benefit by being conspicuously white in the early summer—they do not molt out of their winter white until June—because the cocks, standing on their look-out boulders near the sitting hens, are so obvious that they are captured in "thousands" by birds of prey. Since they are so numerous, their losses are insignificant, while their "sacrifice" preserves the all-important hens.

Finally, why are snow geese breeding at least as far north

as 82° in the Canadian Arctic pure white, but brent geese, probably breeding even further north in Peary Land, black and white? Both species nest on the ground, and the brent are known to be heavily preyed on by foxes. No doubt snow geese are too, and according to Russian sources those breeding on Wrangel Island frequently do so around the nesting hillocks of snowy owls, whose territory no fox will approach. One wonders why the brent geese do not site their nests protectively on the craggy outcrops of hills and mountain walls, as do the black and white barnacle geese which breed as far north as 77 degrees in north-east Greenland.

By and large there is clearly little evidence that whiteness has any significant camouflagic value for either prey or predators, though in summer a polar bear's yellowish-white pelage may render him less conspicuous when stalking basking seals over ice stained with algae; while, in winter, white wolves will be less conspicuous than gray or near-black ones when hunting caribou over the snow. Conversely, though, the fact that the caribou of the High Arctic are pale-colored does not protect them from wolves hunting mainly by ear and nose. If whiteness has a camouflagic value, then it should be more prevalent in the Low Arctic, where the overall snowfall is heavier than in the High Arctic.

An alternative controller of whiteness might be temperature. Not all the collared lemmings inhabiting some of the Alaskan islands, where temperatures are relatively mild, turn white in winter. It may be that Ross demonstrated the affect of temperature conclusively more than 135 years ago, during the course of his voyage in search of the North-West Passage, by the reactions of a collared lemming which he kept in his cabin during the winter. In the warmth of the cabin the lemming retained its dark summer coat. In

February, however, its cage was removed to the deck, and after a week, when the temperature fell to —30 degrees F, the lemming had turned completely white.

As a generalization it can be said that most Arctic birds and mammals become white (or paler in color) in winter. They do so, it is suggested, because by being white they lose less heat from radiation than if they were dark colored; and the prevention of heat-loss is probably of greater importance than the relatively small amount of heat absorbed from the sun by being dark. Ptarmigan, for example, begin to turn white on the underparts as early as August when night temperatures are already falling below freezing, and lemmings and ermines in September. But if whiteness is stimulated by changes in temperature, why are ravens and musk-oxen unaffected, and why does the tundra fox not begin to turn white until November or December, long after the

Ermine in winter coat

freeze-up and permanent snow covering? Those tundra foxes which descend to the coastal environment of the blue foxes in the winter still change into white. Perhaps whiteness is the response to a variety of stimuli, such as changes in temperature or decreasing daylight, rather than the product of natural selection as a protective measure.

On the other hand the dense tight feather-mail of ptarmigan, the long silky hair of hares, and the thick coats of larger mammals are obviously protective winter clothing, irrespective of their color. The feathers of Arctic birds are larger and more numerous, and their downy bases denser and more extended than those of relatives inhabiting more southerly latitudes. Small mammals, such as foxes, hares and lemmings, have woolly fur beneath their silky outer guard-hairs, and in very low temperatures can erect the latter, as birds do their feathers, increasing insulation by more than 100 per cent; while the mat of hair overlaying the dense under-fur of larger mammals, such as musk-oxen and caribou, forms an almost airtight coat. Possibly the compact form and shorter extremities of some Arctic animals, in comparison with those of more southerly kin, are also instrumental in retaining heat, though caribou and wolves hardly fall into this category. The feet and legs of birds and mammals may be feathered or furred. The hair on those of Arctic lemmings and ermines is much denser than that of other races. The sole of the Arctic hare's exceptionally large hind foot, 7 inches long, bears a soft fine-spun fur. Thick hair covers the fox's heel and toe pads, and dense close-set hair the polar bear's soles. Hairy or feathered feet are no doubt protected against the effects of cold, though a bird's feet contain so few blood-cells that their temperature cannot be much above freezing point. But a foot-covering must also be an aid to sure-footedness on ice and snow, and all Arctic animals have relatively large feet acting as snow-shoes. The bear's hairy soles, and strong claws, grip ridges in the frozen snow and the slightest fracture of wind-polished glassy ice with equal efficiency, and he can travel with astonishing agility for such a heavy beast at a pace of 20 miles per hour over the roughest ice, leaping nimbly over 6-foot hummocks and bouncing off the ice like a rubber ball.

To sum up then, there is no one satisfactory explanation of whiteness among polar animals, nor any evidence that it is essential to existence in polar regions. A thick layer of fat, dense fur or feathers, and the ability to obtain food and to move efficiently in the most extreme weather conditions are the fundamental necessities.

In Conclusion

What can one say about the future of the two polar worlds and their inhabitants? Ultimately nothing is sacred to governments. Militarism, power politics, big business are always the final arbiters. They may lose a battle, but they always win the war—or have done so up to now. It is true that all those governments with sectors of influence neatly ruled out on the white map of Antarctica signed in 1961 The Antarctic Treaty Bill for the conservation of its fauna and flora in perpetuity, with the intention that Antarctica is to be preserved for all time as a natural scientific laboratory and meteorological station, where we can learn much about the climate and resources of the Earth. But is this treaty only the traditional worthless scrap of paper, to be torn up by one or other of the signatories if expedience indicates that it is profitable to do so?

The intrinsic values of the treaty are in fact already being eroded. The very establishment and inbeing of large scientific expeditions in an Antarctic environment necessitates constant supplying on an immense scale. While modern transport avoids the earlier heavy exploitation of Weddell seals and Adélie penguins as essential food for men and dogs, it does however require huge and complex bases, ports and airfields, and shuttle-services of military and civil transport and supply ships, spilling oil, which must inevitably introduce pollution of one kind or another and colossal dumps of rubbish which cannot be disposed of in Antarctica's gigantic refrigerator of ice and snow—the only snows in Earth not hitherto contaminated by carbon-monoxide.

Nor can para-military or even scientific personnel always be relied upon to behave responsibly towards the environment and its fauna. The remotest rookeries of emperor penguins, previously inviolate, are now within reach of and subject to disturbance by helicopters and motorized transport. Moreover, a new feature of Antarctica is the tourist, who could turn out to be big business now that there are airfields for his reception and ice-breakers powerful enough to forge a way through the thickest pack-ice. Seals and the colonies of Adélie penguins and other breeding birds must now accustom themselves to batteries of cameras.

Fortunately, Antarctica does not appear to be of any significant military value to twentieth-century strategists, though there has been no concealment of the fact that the USA's sophisticated bases serve as a para-military trying-out ground for men, machinery and equipment under polar conditions; and the capitals of the Southern Ocean lie on a potential trans-antarctic air route. Nor is Antarctica economically exploitable at the present time, for its minerals and vast coalfields are buried beneath several thousand feet of snow and ice, so that the cost of extracting these, even if practicable, are prohibitive. On the other hand the Antarctic's fish stocks have never been tapped, though we have seen what happened to its supposedly inexhaustible reservoir of whales. Between 1904 and 1966 totals in excess of 670,000 fin whales, 330,000 blue whales, 145,000 hump-backed whales and 87,000 sei whales are known to have been killed. The blue whales and hump-backs were virtually exterminated, the fin whales reduced to skeleton herds, and the numbers of sei whales cut by considerably more than half. Happily, it seems possible that some degree of control, albeit strongly opposed, over the catches of the whaling fleets during the past few years may just have come in time to preserve sufficient stocks for the rebuilding of the herds. At present this is no more than a possibility, though the fur seals and elephant seals, which were also almost exterminated by whalers and sealers during the nineteenth and early twentieth centuries, are beginning to re-establish rookeries on the Antarctic islands. Their recovery will be speeded by the decision of the twelve Antarctic Treaty governments to agree to voluntary control of sealing in the pack-ice, and to the establishment of six sealing zones, each of which will be closed in rotation annually. The result is that these seals, and also the Ross seals, are fully protected in the ice-fields, while annual kill quotas have been fixed for crab-eater, leopard and Weddell seals, together with full protection for the latter from March to August. But, learning from the tragedy of the whales, it is clear that any Antarctic fishery would have to operate under the strictest international controls, while any attempt to harvest krill on an economic scale—fortunately, a fairly remote possibility—must be totally resisted. Krill are, as we have seen, the basis of all life in the Antarctic, and large-scale depletion of their myriads could destroy the ecological balance and delicately linked food-chain throughout the Antarctic.

If there appears to be no immediate threat to the wild life of Antarctica, though pollution and disease are always potential threats, the same cannot be said for the Arctic. Its military and economic significance and its long and bloody history of exploration and exploitation have destroyed the cultures, economies and way of life of most Eskimo, Samoyede and Chukchi peoples, though a few have been absorbed into modern fishing industries. Today, military and civil aircraft fly hourly over the North Pole, nuclear submarines pass under its ice, early-warning systems span the Canadian Arctic, and there is a colossal military airbase in north-west Greenland. The USA blasts the sub-Arctic Aleutian Islands with nuclear earthquakes, and the Soviet Union does the same in Novaya Zemlya. The Arctic lichens absorb heavy concentrations of Strontium-90 and Ceasium-137, caribou and reindeer feed on the lichens, and the Lapps who eat the meat and drink the milk of their reindeer herds are exposed to the highest recorded levels of radioactive fall-out. The residues of pesticides poison the fats of seals and sea-birds, and oil-slicks pollute Arctic waters. Atmospheric pollution, with a particularly high lead concentration, at Fairbanks in Alaska is comparable to that of Los Angeles' smog. But the exploitation of oil presents perhaps the greatest threat to the Arctic's environment and remaining wildlife. The laying of oil pipelines will break the essential migratory routes from winter to summer pastures of the caribou, already critically reduced in numbers even though they are a potential meat supply in a hungry world. The erosion ditches and tracks of the oil companies' winter road network in Alaska have devastated the terrain to such an extent that an observer from the Cornell Laboratory of Ornithology has stated that, except in the mountainous fastnesses of the Brooks Range, there may not be a single 100-square-mile plot of land east of the Colville River which has not sustained some irreparable damage, despite the fact that this area includes the Arctic National Wildlife Refuge.

Is there no glimmer of hope for the Arctic? Not unless the common people and the conservationists can control their politicians, military chiefs and business moguls. There seems little likelihood of polar bears surviving in the American Arctic where "civilization" has advanced in strength too far north for them. And in the European Arctic, as on the Alaskan ice, the hunting of adult bears and the trade in cubs, which involves killing the she-bears, is apparently uncontrollable. But the current strict conservation of bears in the Soviet Arctic where many den-up to cub—particularly on Wrangel Island where a musk-ox farm is also being established—may allow a sufficient stock to survive to offset the losses resulting from their nomadic way of life, in which they wander out of protected zones. It is in the Soviet Arctic that walruses, after being subjected to centuries of slaughter on a vast scale, are showing signs of increasing and are expanding their range. Of the great Arctic whales which have been almost exterminated, the hump-backs and gray whales are slowly recovering, thanks to rigorous protection during their bi-annual migrations along the Pacific coast of America. And it is now known that a small stock of right-whales has also survived. Nature has demonstrated again and again marvellous powers of regeneration. But only constant vigilance can maintain even the present precarious ecological status in the Arctic.

Glossary

Aglo	Eskimo term for seal's breathing hole in ice
Amphipod	small shrimp-like crustacean
Bay ice	concentration of ice-floes, often several years old
Fast-ice	sea-ice attached to land
Ice foot	edge of fast-ice
Ice shelf	front edge of glacier or land-ice, often afloat
Bulbil	small bulb or tuber
Echinoderm	a "tube-foot," such as sea-urchin or starfish
Euphausian	small shrimp-like crustacean
Iglooviuk (Igloo)	Eskimo snow-house
Loomery	nesting concentration of auks on cliffs
Nunatak	rock peak protruding above snow
Polynia	pool of open water in pack-ice
Pteropod	minute winged mollusk
Sauggsat	shrinking polynia in ice, in which white whales and narwhals have been trapped
Springtail	minute wingless insect
Ugli (pl. Uglit)	walruses' rookery
Umiak	originally Eskimo women's large skin-boat, rowed with oars
Water-bear	8-legged mite-like arthropod

Bibliography

Key to abbreviations:
ANARE—*Australian Antarctic Research Expedition*
BGLE—*The British Graham Land Expedition, 1934–1937*
FIDS—*Falkland Islands Dependencies Survey: Scientific Reports*
MOIPBOB—*Moskovskoe Obshchestvo Ispytatelei Prirody, Biulleten, Otdel Biologicheskii*
ZSL—*Zoological Society, London*

The Antarctic

BOOKS

Amundsen, Roald E. *The South Pole: An Account of the Norwegian Antarctic Expedition in the "Fram," 1910–1912*. Trans. by A. G. Chater. New York: Keedick, 1913.
Anderson, William E. *Expedition South*. London: Evans, 1960.
Bagshawe, Thomas W. *Two Men in the Antarctic: An Expedition to Graham Land, 1920–1922*. New York: Macmillan, 1939.
Bernacchi, Louis C. *To The South Polar Regions: Expedition of 1898–1900*. London: Hurst & Blackett, 1901.
Bertram, George C. *The Biology of the Weddell and Crab-Eater Seals*. BGLE, vol. 1, no. 1. London: The British Museum, 1940.
Billing, Graham, and Mannering, Guy. *South: Man and Nature in Antarctica*. Seattle: University of Washington Press, 1964.
Borchgrevink, Carsten E. *First on the Antarctic Continent: Being an Account of the British Antarctic Expedition, 1898–1900*. London: Newness, 1901.
Brown, D. A. *Breeding Biology of the Snow Petrel*. ANARE: *Scientific Reports*, series B, vol. 1. Melbourne, 1966.
Brown, K. G. *The Leopard Seal at Heard Island, 1951–1954*. ANARE: *Interim Reports*, no. 16. Melbourne, 1957.
Bruce, William S. *Polar Exploration*. New York: Holt, 1911.
Bursey, Jack. *Antarctic Night: One Man's Story of 28,224 Hours at the Bottom of the World*. New York: Rand McNally, 1957.

Cherry-Garrard, Apsley G. *The Worst Journey in the World.* New York: MacVeagh, Dial, 1930.
Coleman-Cooks, John. *Discovery II in the Antarctic: The Story of British Research in the Southern Seas.* London: Odhams, 1963.
Dröscher, Vitus B. *The Magic of Senses: New Discoveries in Animal Perception.* Trans. by Ursula Lehrburger and Oliver Coburn. New York: Dutton, 1969.
Fuchs, Vivian, and Hillary, Edmund. *The Crossing of Antarctica: The Commonwealth Trans-Antarctic Expedition, 1955–1958.* London: Cassell, 1959.
Gillham, Mary E. *Sub-Antarctic Sanctuary: Summertime on Macquarie Island.* London: Gollancz, 1967.
Glenister, T. W. *The Emperor Penguins.* Vol. 2, *Embryology.* FIDS, no. 10. London: H.M. Stationery Office, 1954.
Hardy, Alister C. *Great Waters: A Voyage of Natural History to Study Whales, Plankton and the Waters of the Southern Ocean.* New York: Harper & Row, 1967.
Hatherton, Trevor, ed. *Antarctica.* New York: Praeger, 1965.
Herbert, Wally. *A World of Men: Exploration in Antarctica.* New York: Putnam's, 1969.
Holdgate, Martin W., ed. *Antarctic Ecology.* New York: Academic 1970.
Huxley, L., ed. *Scott's Last Expedition.* New York: Dodd, Mead, 1913.
King, H. G. R. *The Antarctic.* London: Blandford, 1969.
King, Judith E. *Seals of the World.* London: The British Museum, 1964.
Kirwan, L. P. *The White Road: A Survey of Polar Exploration.* London: Hollis & Carter, 1959.
Laws, R. M. *The Elephant Seal.* FIDS, no. 8. London: H.M. Stationery Office, 1953.
Levick, George M. *Antarctic Penguins: A Study of Their Social Habits.* London: Heinemann, 1914.
Lewis, Richard S. *A Continent for Science: The Antarctic Adventure.* New York: Viking, 1965.
Lillie, Harry R. *The Path Through Penguin City.* London: Benn, 1965.
Mansfield, Arthur W. *Breeding Behavior and the Reproductive Cycle of the Weddell Seal.* FIDS, no. 18. London: H.M. Stationery Office, 1958.
Marret, Mario. *Seven Men Among the Penguins: An Antarctic Ven-*

ture. Trans. by Edward Fitzgerald. New York: Harcourt, Brace, 1955.

Matthews, Leonard H. *Sea Elephant: The Life and Death of the Elephant Seal*. London: MacGibbon & Kee, 1952.

———. *South Georgia: The British Empire's Sub-Antarctic Outpost*. London: Wright, 1931.

———. *Wandering Albatross: Adventures Among the Albatrosses and Petrels in the Southern Ocean*. London: MacGibbon & Kee, 1951.

Mawson, Douglas. *The Home of the Blizzard: Being the Story of the Australasian Antarctic Expedition, 1911–1914*. Philadelphia: Lippincott, 1915.

Maxwell, Gavin. *Seals of the World*. Boston: Houghton Mifflin, 1967.

Murphy, Robert C. *Logbook for Grace: Whaling Brig Daisy*. New York: Macmillan, 1947.

———. *Oceanic Birds of South America*. 2 vols. New York: Macmillan and The American Museum of Natural History, 1948.

Ommanney, Francis D. *South Latitude*. New York and London: Longmans, Green, 1938.

Perry, Richard. *The Unknown Ocean*. New York: Taplinger, 1972.

Ponting, Herbert G. *The Great White South: Or, With Scott in the Antarctic*. New York: McBride, 1923.

Prévost, Jean. *Écologie du Manchot Empéreur*. Expéditions Palaires Françaises, publ. 222. Paris: Hermann, 1961.

Priestley, Raymond; Adie, Raymond J.; and Robin, G. de Q., eds. *Antarctic Research: A Review of British Scientific Achievements in Antarctica*. London: Butterworths, 1964.

Rankin, Niall. *Antarctic Isle: Wild Life in South Georgia*. London: Collins, 1951.

Reader's Digest Association. *Living World of Animals*. New York: Reader's Digest Association, 1970.

Rivolier, Jean. *Emperor Penguins*. Trans. by Peter Wiles. London: Elek, 1956.

Rofen, Robert R., and Dewitt, H. H. *Antarctic Fishes*. *Science in Antarctica*, part 1. Committee on Polar Research, National Research Council, publ. 839. Washington, D.C.: National Academy of Science, 1961.

Rymill, John. *Southern Lights: The Official Account of the British Graham Land Expedition*. London: Chatto & Windus: 1938.

Saunders, Alfred. *A Camera in Antarctica*. London: Winchester, 1950.

Scott, Robert F. *The Voyage of the "Discovery."* New York: Scribner's, 1905.

Shackleton, Ernest H. *The Heart of the Antarctic: Being the Story of the British Antarctic Expedition, 1907–1909.* Philadelphia: Lippincott, 1909.

Sparks, John, and Soper, Tony. *Penguins.* New York: Taplinger, 1967.

Stonehouse, Bernard. *The Brown Skua of South Georgia.* FIDS, no. 14. London, H.M. Stationery Office, 1956.

———. *The Emperor Penguin.* Vol. 1, *Breeding, Behaviour and Development.* FIDS, no. 6. London: H.M. Stationery Office, 1953.

———. *The General Biology and Thermal Balance of Penguins. Advances in Ecological Research,* ed. by J. B. Cragg, vol. 4. New York: Academic, 1967.

———. *The King Penguin of South Georgia.* Vol. 1, *Breeding, Behaviour and Development.* FIDS, no. 23. London: H.M. Stationery Office, 1960.

———. *Penguins.* New York: Golden, 1968.

Sullivan, W. *Quest for a Continent.* New York: McGraw-Hill, 1957.

Wilson, Edward A. *Aves. National Antarctic Expedition, 1901–1904,* vol. 2, part 3. London: The British Museum, 1907.

———. *Birds of the Antarctic.* Ed. by Brian Roberts. New York: Humanities, 1968.

———. *Diary of the "Discovery" Expedition to the Antarctic.* Ed. by Ann Savours. New York: Humanities, 1967.

———. *Mammalia. National Antarctic Expedition, 1901–1904,* vol. 2, part 1. London: The British Museum, 1907.

ARTICLES

Budd, G. M. "The Biotopes of Emperor Penguin Rookeries." *Emu* 61 (1961): 171–189.

———. "Population Studies in Rookeries of the Emperor Penguin." *Proc ZSL* 139 (1962): 365–368.

———. "Sledging to the Emperors." *Animals* 4 (1964): 390–395.

Cameron, A. S. "The Auster Emperor Penguin Rookery, 1965." *Emu* 69 (1969): 103–106.

Caughley, Graeme. "The Cape Crozier Emperor Penguin Rookery." *Dominion Museum of Wellington, New Zealand* 3 (1960): 251–262.

Cloudsley-Thompson, John L. "Animal Life in Snow and Ice." *Animals* 5 (1964): 178–187.

Cranfield, H. J. "Emperor Penguin Rookeries of Victoria Land." *Antarctic* (Wellington) 4 (1966): 365–366.

Eklund, Carl R. "Distribution and Life-History Studies of the South-Polar Skua." *Bird-Banding* 32 (1961): 187–223.

Emlen, J. T., and Penney, R. L. "The Navigation of Penguins." *Scientific American* 215 (1966): 104–113.

Guillard, Robert, and Prévost, Jean. "Observations Écologiques à la Colonie des Manchots Empéreurs de Pointe Géologie (Terre Adélie) en 1963." *L'oiseau et la Revue Française d'Ornithologie* 34 (1964): 33–51.

Hamilton, J. E. "The Leopard Seal." *Discovery Reports* 18 (1939): 239–264.

Harrington, H. J. "Narrative of a Visit to the Newly Discovered Emperor Penguin Rookery at Coulman Island, Ross Sea." *Notornis* 8 (1959): 127–132.

Laws, R. M., and Taylor, R. J. F. "A Mass Dying of Crab-Eater Seals." *Proc* ZSL 129 (1957): 315–324.

Prévost, Jean. "How Emperor Penguins Survive the Antarctic Climate." *New Scientist* 16 (1962): 444–447.

Pryor, Madison E. "Brood Mortality of the Offspring of Emperor Penguins." *Problemy Arktici i Antarktici, Vypusk* 19 (1965): 59–61. (Trans. by National Lending Library for Science and Technology, Boston, 1971.)

Ray, Carleton, and Schevill, William E. "Noisy Underworld of the Weddell Seal." *Animals* 10 (1967): 109–113.

Rivolier, Jean. "Polar Realm of the Emperors." *Natural History* 68 (1959): 66–81.

Sapin-Jaloustre, Jean. "Découverte et Description de la Rookery de Manchot Empéreur de Pointe Géologie." *L'oiseau et la Revue Française d'Ornithologie* 22 (1952): 225–260.

Stonehouse, Bernard. "Emperor Penguin Colony at Beaufort Island, Ross Sea." *Nature* 210 (1966): 925–926.

———. "Emperor Penguins at Cape Crozier." *Nature* 203 (1964): 849–851.

Taylor, Robert J. F. "Three Species of Whale Being Restricted to Pools in the Antarctic Sea-Ice." *Proc* ZSL 129 (1957): 325–331.

Willing, Richard L. "Australian Discoveries of Emperor Penguin Rookeries in Antarctica During 1954–1957." *Nature* 182 (1958): 1393–1394.

Young, E. C. "Breeding Behaviour of South Polar Skua." *Ibis* 105 (1963): 202–233.
———. "Feeding Habits of South Polar Skua." *Ibis* 105 (1963): 1310–1318.

The Arctic

BOOKS

Amundsen, Roald E. *My Polar Flight.* London: Hutchinson, 1925.
———. *The North-West Passage: Being the Record of a Voyage of Exploration of the Ship "Gjoa," 1903–1907.* New York: Dutton, 1908.
Anderson, Rudolph M. "Mammals of the Eastern Arctic and Hudson Bay" and "Arctic Flora." In *Canada's Eastern Arctic: Its History, Resources, Population and Administration,* assembled by W. C. Bethune, pp. 67–108, 133–137. Ottawa: Pateraude, 1935.
Baĭdukov, Georgiĭ E. *Over the North Pole.* Trans. by Jessica Smith. New York: Harcourt, Brace, 1938.
Baird, Patrick. *The Polar World.* New York: Wiley, 1964.
Bannerman, David. *Birds of the British Isles,* vol. 11. Edinburgh: Oliver & Boyd, 1962.
Bartlett, Robert A. *The Last Voyage of the Karluke.* Comp. by Robert T. Hale. Boston: Small, Maynard, 1916.
Beechey, Frederick W. *Narrative of a Voyage to the Pacific and Bering's Strait to Co-operate with the Polar Expeditions.* Philadelphia: Carey & Lea, 1832.
Belcher, Edward. *The Last Arctic Voyages: Being a Narrative of the Expedition . . . in Search of Sir John Franklin.* London: Reeve, 1855.
Berton, Pierre. *The Mysterious North.* New York: Knopf, 1956.
Biggar, H. P. *Voyages of Jacques Cartier.* The Public Archives of Canada, publ. 2. Ottawa: Acland, 1924.
Böcher, Tyge W.; Holmen, Kjeld; and Jakobsen, Knud. *The Flora of Greenland.* Trans. by T. T. Elkington and M. C. Lewis. Copenhagen: Hasse, 1968.
Brooks, James W. *A Contribution to the Life History and Ecology of the Pacific Walrus.* Alaska Cooperative Wildlife Research Unit, spec. rpt. 1. Juneau, 1954.
———. "The Management and Status of Marine Mammals in Alaska." In *Transactions of the 28th North American Wildlife*

and Natural Resources Conference, March 4, 5, 6, 1963. Washington, D.C., 1963.

Buckley, John L. *The Pacific Walrus: A Review of Current Knowledge and Suggested Management Needs.* U.S. Fish and Wildlife Service, spec. science rpt.—wildlife 41. Washington, D.C., 1958.

Burns, John J. *The Walrus in Alaska: Its Ecology and Management. Federal Aid in Wildlife Restoration Project: Reports,* vol. 5. Juneau: Alaska Department of Fish and Game, 1965.

Cabot, William B. *In Northern Labrador.* Boston: Badger, 1912.

Cahalane, Victor H. *Mammals of North America.* New York: Macmillan, 1947.

Carrighar, Sally. *Icebound Summer.* New York: Knopf, 1953.

Chapman, F. S. *Northern Lights: The Official Account of the British Arctic Air-Route Expedition.* New York: Oxford University Press, 1934.

Chapsky, K. K. *The Walrus of the Kara Sea: Results of the Investigation of the Life History, the Geographical Distribution and Stock of Walruses in the Kara Sea and Novaya Zemlya.* Leningrad: Vsesoivznyĭ Arkticheskiĭ Institut, 1936. (Summary in English.)

Conway, William M., ed. *Early Dutch and English Voyagers to Spitsbergen in the 17th Century, Including Hessel Gerritsz "Historie du Pays Nommé Spitsberghe," 1613 . . . and Jacob Segersz van der Brugge "Journal of Dagh Register,"* Amsterdam, 1634. Trans. by Basil H. Soulsby and J. A. F. De Williers. *Works,* series 2, no. 11. London: The Hakluyt Society, 1904.

———. *No Man's Land: History of Spitsbergen from Its Discovery in 1596 to the Beginning of the Scientific Exploration of the Country.* Cambridge: University Press, 1906.

Cook, Frederick. *My Attainment of the Pole: Being the Record of the Expedition that First Reached the Boreal Center, 1907–1909, with the Final Summary of the Polar Controversy.* New York: Polar, 1911.

Cook, James, *A Voyage to the Pacific Ocean.* London: Hughes, 1785.

Courtauld, Augustine, comp. *From the Ends of the Earth: An Anthology of Polar Writings.* New York: Oxford University Press, 1958.

Darling, Frank F. *Pelican in the Wilderness: Or, A Naturalist's Odyssey in North America.* New York: Random House, 1956.

Downes, Prentice G. *The Sleeping Island: The Story of One Man's Travels in the Great Barrier Lands of the Canadian North.* New York: Coward-McCann, 1943.

Dunbar, Maxwell J. *Ecological Developments in Polar Regions: A Study in Evolution.* Englewood Cliffs, N.J.: Prentice-Hall, 1968.

Egede, Hans. *A Description of Greenland.* Trans. from the Danish. London: Hitch, 1745.

Elliot, Henry W. *An Arctic Province: Alaska and the Sea Islands.* London: Sampson, Low, Marston, Searle & Rivington, 1886.

———. "Report on the Seal Islands of Alaska." In *The 10th Census of the United States,* 1880, vol. 8, pp. 3–188. Washington, D.C.: Government Printing Office, 1884.

Elton, Charles S. *Voles, Mice and Lemmings: Problems in Population Dynamics.* Oxford: Clarendon, 1942.

Erickson, Albert W. *Bear Investigations,* work plan F. Juneau: Alaska Department of Fish and Game, 1961.

Euller, John. *Arctic World.* New York: Abelard-Schuman, 1958.

Fabricius, Otto. *Fauna Groenlandica, Systematice Sistems Animalia Groenlandiae Occidentalis Hactenus Indigata . . . Maximaque Parte Secundum Proprias Observationes Othonis Fabricii.* Copenhagen and Leipzig: Rothe, 1780.

Fay, Francis H. "History and Present Status of the Pacific Walrus Population." In *Transactions of the 22nd North American Wildlife and Natural Resources Conference,* pp. 431–445. Washington, D.C., 1957.

———. "The Pacific Walrus: Spatial Ecology, Life History and Population." Ph.D. thesis, University of British Columbia, 1955.

Fisher, James, and Lockley, Ronald M. *Sea-Birds: An Introduction to the Natural History of the Sea-Birds of North America.* Boston: Houghton Mifflin, 1954.

Formozov, A. N. *Snow Cover as an Integral Factor of the Environment.* Boreal Institute, occasional paper 1. Edmonton: University of Alberta Press, 1964.

Freuchen, Peter. *Arctic Adventure: My Life in the Frozen North.* New York: Farrar & Rinehart, 1936.

———. *Vagrant Viking: My Life and Adventures.* Trans. by Johan Hamkro. New York: Messner, 1953.

———, and Degerkøl, Magnus. *Thule Expedition, 5th, 1921–1924: Report.* Copenhagen: n.p., 1940.

———, and Salomensen, Finn. *The Arctic Year.* New York: Putnam's. 1958.

Gillsäter, Sven. *Wave After Wave.* Trans. by Michael Heron. London: Allen & Unwin, 1964.

Grierson, John. *Challenge to the Pole: Highlights of Arctic and Antarctic Aviation.* Hamden Court: Archon, 1964.

Hagenbeck, Carl. *Beasts and Men: Being Carl Hagenbeck's Experiences for Half a Century Among Wild Animals.* Trans. by Hugh S. R. Elliot and A. G. Thacker. London and New York: Longmans, Green, 1912.

Haig-Thomas, David. *Tracks in the Snow.* London: Hodder & Stoughton, 1939.

Hall, Charles F. *Arctic Researches and Life Among the Esquimeau: Being the Narrative of an Expedition in Search of Sir John Franklin.* New York: Harper, 1865.

Hanbury, David T. *Sport and Travel in the Northland of Canada.* New York: Macmillan, 1904.

Harington, C. Richard. *Denning Habits of the Polar Bear.* Canadian Wildlife Service Report Series, no. 5. Ottawa, 1968.

———. *Some Data on the Polar Bear and Its Utilization in the Canadian Arctic.* Ottawa: Canadian Wildlife Service, 1961.

Hearne, Samuel A. *A Journey from the Prince of Wales' Fort in Hudson's Bay to the Northern Ocean.* London: Strahan and Cadell, 1795.

Herbert, Wally. *Across the Top of the World; The British Trans-Arctic Expedition.* London: Longmans, Green, 1969.

Hewitt, Charles G. *The Conservation of the Wildlife of Canada.* New York: Scribner's, 1921.

Howell, Alfred B. *Aquatic Mammals.* Baltimore: Thomas, 1930.

Illingworth, Frank. *Wild Life Beyond the North.* London: Country Life, 1951.

Jackson, Frederick G. *A Thousand Days in the Arctic.* New York: Harper, 1899.

Jensen, Adolf S. *Grønlands Fauna.* Copenhagen: Lunesbogty Kkeri, 1928.

Kane, Elisha. *Arctic Explorations in Search of Sir John Franklin.* Philadelphia: Childs & Petersen, 1856.

Kimble, George H., and Good, Dorothy, eds. *Geography of the Northlands.* New York: American Geographical Society and Wiley, 1955.

Koch, Lauge. *Au Nord du Groenland.* Paris: Roger, 1928.

Kumlien, L. *Contributions to the Natural History of Arctic America Made in Connection with the Howgate Polar Expedition, 1877–1878.* U.S. National Museum bull. 13; Smithsonian Institution publ. 342. Washington, D.C.: Government Printing Office, 1879.

Lamont, James. *Seasons with the Sea-Horses: Or, Sporting Adventure in the Northern Seas*. London: Hurst & Blackett, 1861.

Lønø, Odd. *The Catches in Polar Bears in Arctic Regions in the Period 1945–1963*. Arbok 1963. Oslo: Norsk Polarinstittut, 1965.

———. *Some Remarks on Polar Bear Biology*. Oslo: n.p., 1962.

Loughrey, Alan G. *Preliminary Investigation of the Atlantic Walrus*. Canadian Wildlife Service wildlife management bull. series 1, no. 14. Ottawa, 1959.

Low, Albert P. *Report on the Dominion Government Expedition to Hudsons Bay and the Arctic Islands on Board the D. G. S. Neptune, 1903–1904*. Ottawa: Government Printing Bureau, 1906.

McClintock, Francis L. *Voyage of the Fox in the Arctic Seas: Narrative of the Discovery of the Fate of Sir John Franklin*. Boston: Ticknor & Fields, 1860.

MacMillan, Donald B. *Etah and Beyond: Or, Life Within Twelve Degrees of the Pole*. Boston and New York: Houghton Mifflin, 1927.

———. *Four Years in the White North*. New York and London: Harper, 1918.

Malaurie, J. *The Last Kings of Thule: A Year Among the Polar Eskimos of Greenland*. Trans. by Gwendolen Freeman. New York: Crowell, 1956.

Manniche, Arnor L. V. *Terrestrial Mammals und Birds of North-East Greenland*. Meddelelser om Grønland, ed. by Martin Vahl, vol. 45. Copenhagen: Reitzel, 1928.

Mansfield, Arthur W. *The Biology of the Atlantic Walrus in the Eastern Canadian Arctic*. Montreal: Fisheries Research Board of Canada Arctic Unit, 1958.

Marsden, Walter. *The Lemming Year*. London: Chatto & Windus, 1964.

Mikkelsen, Ejnar. *Conquering the Arctic Ice*. London: Heinemann, 1909.

———. *Lost in the Arctic: Being the Story of the "Alabama" Expedition, 1909–1912*. New York: Doran, 1913.

Mirsky, J. *To the Arctic! The Story of Northern Exploration from the Earliest Times to the Present*. New York: Knopf, 1948.

Montague, Sydney R. *North to Adventure*. New York: McBride, 1939.

Munn, Henry T. *Prairie Trails and Arctic By-Ways*. London: Hurst & Blackett, 1932.

Nansen, Fridjof. *Farthest North: Being a Record of a Voyage of Exploration of the Ship "Fram," 1893–1896*. New York: Harper, 1898.

———. *The First Crossing of Greenland.* Trans. by Hubert M. Gepp. London and New York: Longmans, Green, 1890.

———. *Hunting and Adventure in the Arctic.* New York: Duffield, 1925.

———. *In Northern Mists: Arctic Exploration in Early Times.* Trans. by Arthur Chater. New York: Stokes, 1911.

Nares, George S. *Narrative of a Voyage to the Polar Sea.* Ed. by Henry W. Feilden. London: Low, Marston, Searle & Rivington, 1878.

Nelson, Edward W. *Wild Animals of North America.* Washington, D.C.: The National Geographic Society, 1930.

Noice, Harold. *With Stefansson in the Arctic.* New York: Dodd, Mead, 1924.

Novkov, Georgiĭ A. *Carnivorous Mammals of the Fauna of the U.S.S.R.* Trans. by A. Birron and Z. S. Cole. Jerusalem: Israel Program for Scientific Translations, 1962. (Available from Off. of Technical Services, U.S. Dept. of Commerce, Washington, D.C.)

Ognev, Sergei I. *The Mammals of the U.S.S.R. and Adjacent Countries.* 9 vols. Trans. by A. Birron and Z. S. Cole. Published for the National Science Foundation by the Israel Program for Scientific Translations, 1962–1967.

Orléans, Louis Philippe Robert, duc d'. *Hunters and Hunting in the Arctic.* Trans. by H. Grahame Richards. London: Nutt, 1911.

Papinin, Ivan D. *Life on an Ice Floe.* Trans. by Fanny Smitham. New York: Hutchinson, 1947.

Peary, Robert E. *Nearest to the Pole.* New York: Doubleday, Page, 1907.

———. *The North Pole.* New York: Stokes, 1910.

———. *Northward over the Great Ice.* New York: Stokes, 1898.

Pedersen, Alwin. *Der Eisbär: Verbreitung und Lebensweise.* Copenhagen: Bruun, 1945.

———. *Polar Animals.* Trans. by Gwynne Vevers. New York: Taplinger, 1966.

———. *Rosmarus.* Copenhagen: Gyldendal, 1951.

———. *Das Walross.* Wittenberg Luthertstadt: Ziemsen, 1962.

Perry, Richard. *Bears.* New York: Arco, 1970.

———. *Shetland Sanctuary: Birds on the Isle of Noss.* London: Faber & Faber, 1948.

———. *The World of the Polar Bear.* Seattle: University of Washington Press, 1966.

———. *The World of the Walrus.* New York: Taplinger, 1967.

Preble, Edward A. *Birds and Mammals of the Pribilof Islands, Alaska.*

North American Fauna, vol. 46. Washington, D.C.: Government Printing Office, 1923.

Pruitt, William O. *Animals of the North*. New York: Harper & Row, 1967.

Rasmussen, Knud J. *Across Arctic America: Narrative of the Fifth Thule Expedition*. 1927. Reprint. New York: Greenwood, 1969.

———. *Greenland by the Polar Sea: The Story of the Thule Expedition*. Trans. by Asta and Rowland Kenney. London: Heinemann, 1921.

Ross, James C. *A Voyage of Discovery and Research in the Southern and Antarctic Regions*. London: Murray, 1847.

Ross, John. *Narrative of the Second Voyage in Search of a Northwest Passage . . . Including the Reports of . . . James Clark and the Discovery of the North Magnetic Pole*. London: Webster, 1835.

Scheffer, Victor B. *Seals, Sea-Lions and Walruses: A Review of the Pinnipedia*. Stanford, Calif.: University Press, 1958.

Schwatka, Frederick W. *Nimrod in the North: Or, Hunting and Fishing Adventures in the Arctic Regions*. New York: Cassell, 1885.

Scoresby, William. *Arctic Regions: With a History and Description of the Northern Whale-Fishery*. Edinburgh: Constable, 1820.

———. *Journal of a Voyage to the Northern Whale Fishery: Including Researches and Discoveries on the Eastern Coast of West Greenland*. Edinburgh: Constable, 1823.

Scott, Robert F.; Kenyon, K. W.; Buckley, J. L.; and Olson, S. T. "Status and Management of the Polar Bear and the Pacific Walrus." *In Transactions of the 24th North American Wildlife and Natural Resources Conference, March, 1959*, pp. 366–374. Washington, D.C.: 1960.

Seton, Ernest T. *Lives of Game Animals*. Garden City, N.Y.: Doubleday, Page, 1925–1928.

Simpson, C. J. *North Ice: The British North Greenland Expedition*. London: Hodder & Stoughton, 1957.

Sivertsen, Erling. *On the Biology of the Harp Seal: Investigations Carried Out in the White Sea, 1923–1937*. Hvalradets Skrifter, no. 26. Oslo: Dybwad, 1941.

Soper, Joseph D. *Faunal Investigations of Southern Baffin Islands*. National Museum of Canada Department of Mines biological series 15, bull. 53. Ottawa: Acland, 1928.

Staib, Bjørn O. *On Skis Towards the North Pole*. Trans. by Christopher Nordman. Garden City, N.Y.: Doubleday, 1965.

Standfield, R.; Kolenosky, G.; Shannon, J.; and MacFie, J. *Aerial Surveys of Polar Bear Populations in Ontario.* Wildlife Section, Research Branch, Department of Lands and Forest Reports. Ottawa: 1963.

Stefansson, Vilhjalmur. *The Friendly Arctic: The Story of Five Years in Polar Regions.* New York: Macmillan, 1953.

———. *Hunters of the Great North.* New York: Harcourt, Brace, 1922.

———. *My Life with the Eskimos.* New York: Macmillan, 1929.

———. *Not by Bread Alone.* New York: Macmillan, 1929.

———, and Knight, John J. *The Adventure of Wrangel Island.* New York: Macmillan, 1925.

Sutton, George M., and Hamilton, W. J. *The Mammals of Southampton Island. Carnegie Museum Memoirs,* vol. 12, part 2, sec. 1. Pittsburgh: Carnegie Institute, 1932.

Sverdrup, Otto. *Arctic Adventures: Adapted from New Land.* Trans. by E. H. Hearn and ed. by T. C. Fairley. London: Longmans, Green, 1959.

———. *New Land: Four Years in the Arctic Regions.* Trans. by E. H. Hearn. New York and London: Longmans, Green, 1904.

Uspensky, Savva M. *The Bird Bazaars of Novaya Zemlya.* Translations of Russian Game Reports, vol. 4. Ottawa: Canadian Wildlife Service, 1956.

Veer, Gerrit de. *The Three Voyages of Will Barentz to the Arctic Regions.* London: The Hakluyt Society, 1876.

Vibe, Christian. *Arctic Animals in Relation to Climatic Fluctuations.* Meddelelser om Grønland, ed. by Martin Vahl, vol. 170, no. 5. Copenhagen: Reitzel, 1967.

———. *The Marine Mammals and the Marine Fauna of the Thule District (Northwest Greenland).* Meddelelser om Grønland, ed. by Martin Vohl, vol. 150, no. 6. Copenhagen: Reitzel, 1950.

Whitney, Casper. *On Snow-Shoes to the Barren Grounds.* New York: Harper, 1896.

Wilkinson, Doug. *Land of the Long Day.* New York: Holt, 1956.

ARTICLES

Allen, Glover M. "The Walrus in New England," *Journal of Mammalogy* 11 (1930): 139–145.

Belopol'kii, Lov O. "On the Migration and Ecology of the Pacific Walrus." *Zoologicheskiĭ Zhurnal* 18 (1939): 762–768. (Summary in English. Also trans. by Fisheries Research Board of Canada.)

Bernard, Joseph F. "Walrus Protection in Alaska." *Journal of Mammalogy* 6 (1925): 100–102.

Collins, G. "Habits of the Pacific Walrus." *Journal of Mammalogy* 21 (1940): 138–144.

Fedoseev, G. A. "Concerning the Reserves and Distribution of the Pacific Walrus." *Zoologicheskiĭ Zhurnal* 41 (1962): 1083–1089. (Summary in English. Also trans. by National Lending Library for Science and Technology, Boston, 1966.)

Gillsäter, Sven. "Walrus Summer." *Animals* 6 (1965): 212–217.

Gray, Robert. "Notes on a Voyage to the Greenland Seas in 1888." *Zoologist* 13 (1889): 1–12, 45–51, 95–104.

Gray, R. W. "The Walrus." *The Naturalist* 991 (1939): 201–207.

Haig-Thomas, David. "Polar Bears." *Zoo Life* 2 (1956): 106–109.

Hanna, G. Dallas. "Mammals of the St. Matthew Islands, Bering Sea." *Journal of Mammalogy* 1 (1920): 118–122.

———. "Rare Mammals of the Pribilof Islands, Alaska." *Journal of Mammalogy* 4 (1923): 209–215.

Harington, C. Richard. "Polar Bears and Their Present Status." *Canadian Audubon Magazine* 26 (1964): 3–10.

Hinds, Magery. "Lemmings in Arctic Canada." *Animals* 13 (1971): 764–765.

Lewis, Harrison, F., and Doutt, J. Kenneth. "Records of the Atlantic Walrus and the Polar Bear in the Northern Part of the Gulf of the St. Lawrence." *Journal of Mammalogy* 23 (1942): 365–375.

MacFie, J. A. "Polar Bear in Ontario." *Ontario Fish and Wildlife Review* 1 (1962).

Mansfield, A. W. "The Walrus in the Canadian Arctic." *Canadian Geographical Journal* 72 (1966): 88–95.

Nikulin, P. G. "Chukchee Walrus." *Tikhookeanskiĭ Nauchno-Issledovatel'skii Institut Rybno go Khozia Istua Okean Ografia* 20 (1940): 21–59. (Trans. by University of Alaska.)

Popov, L. A. "Materials on the Reproduction of Walrus in the Laptev Sea." *MOIPBOB* 65 (1960): 25–30. (Summary in English. Also trans. by National Lending Library for Science and Technology, Boston, 1966.)

Robinson, H. W. "The Walrus in British Waters." *North-Western Naturalist* 2 (1936): 155.

Schiller, E. L. "Unusual Walrus Mortality on St. Lawrence, Alaska." *Journal of Mammalogy* 35 (1954): 203–210.

Shuldham, Molyneux. "Account of the Sea-Cow and the Use Made of It." *Philosophical Transactions of the Royal Society of London* 65 (1775).

Sverdrup, Otto. "Second Norwegian Polar Expedition." *Scottish Geographical Magazine* 19 (1903): 337ff.

Tsalkin, V. I. "Materials on the Biology of the Walrus on the Franz Josef Archipelago." *MOIPBOB* 46 (1937): 43–51. (Summary in English. Also trans. by National Lending Library for Science and Technology, Boston, 1966.

Uspensky, Savva M., and Chernyavsky, F. "Distribution, Number and Protection of the White Bear in the Arctic." *MOIPBOB* 70 (1965): 18–24. (Summary in English. Also trans. by National Lending Library for Science and Technology, Boston, 1966.)

———. "Winter Quarters of Polar Bears on Wrangel Island." *MOIPBOB* 54 (1956): 81–86. (Summary in English. Private translation used.)

Index

(*Page numbers in italic indicate illustrations*)